The Makarov Pistol
Volume 1

Soviet Union & East Germany

Henry C. Brown & Cameron S. White

About The Authors

Henry C. Brown Jr was born and raised in North Carolina. He holds a B.S. from East Carolina University in History, a M.A. from Appalachian State University in Education and an A.S. from Wayne Community College. Married to his college sweetheart and father of three lovely daughters, he spends his days working in his wife's dental practice. As a certifiable history fanatic he has travelled in the Middle East, Africa, Europe, and North America visiting historic sites from the Bronze Age to the Second World War. Mr. Brown is a long-time member of the 6th NC American Revolution re-enactment group, a friend of Fort Ticonderoga, member of the Buffalo Bill Museum, a member of Colonial Williamsburg's Duke of Gloucester Society, and a volunteer at Historic Halifax NC State Historic Site.

Cameron S. White, Ph.D. grew up in central Kansas, but he has spent most of his adult life living in Nebraska. After receiving a B.A. in Psychology, *Summa Cum Laude* from the University of Kansas, he went on to earn a M.A. and Ph.D. in clinical psychology from the University of Nebraska. Currently Dr. White lives in Lincoln, Nebraska where he works as a private practice psychologist. Dr. White has always been a student of firearms, especially 20th century semi-automatic military pistols. He is especially interested in Makarov and Tokarev pistols. His other interests include history and American literature. Dr. White writes about the firearms he has studied when he is not working or spending time with his wife and three children.

About Firearms Safety

This book is concerned solely with documenting the history of the Soviet and East German Makarov pistols, and documenting aspects of particular interest to the modern collector.

This book is not intended to be an operator's manual for the modern collector or shooter of any Makarov pistol. Any procedures for maintenance, disassembly and assembly of firearms, procedures for firing firearms, and firearms and cartridge data, are reproduced for the purpose of collector's interest only. They are not intended to be instructional or to be used in any way, and they are neither endorsed nor recommended by the authors and publisher.

Collectors and shooters should observe and comply with their local firearms laws in regard to safety, training, possession, handling and storage of firearms and ammunition.

The authors and publisher will not be held responsible for any injury, death or damage incurred in the proper or improper storage, handing or use of any firearm or ammunition, or improper use of any information or data in this book, intended for collector's interest only.

The Makarov Pistol

Volume 1

Soviet Union & East Germany

Henry C. Brown & Cameron S. White

Edwin H. Lowe Publishing
劉益謙出版社

www.edwinhlowepublishing.com

National Library of Australia Cataloguing-in-Publication entry

Brown, Henry C., author.
The Makarov pistol : Soviet Union & East Germany / Henry C. Brown, Cameron S. White.
9780994168238 (paperback)
Includes bibliographical references and index.
Makarov automatic pistol.
Pistols--Russia (Federation)--History.
Pistols--Germany (East)--History.
Pistols--Identification--Pictorial works.
Pistols--Design and construction.
Pistols--Collectors and collecting.
White, Cameron S., author.
683.4325

Edwin H. Lowe Publishing, Sydney.
www.edwinhlowepublishing.com
contact@edwinhlowepublishing.com
Edwin H. Lowe Publishing is the trading name of Edwin Hulme Lowe. ABN 60 901 995 995

ISBN-13: 978-0-9941682-3-8
ISBN-10: 0-9941682-3-3

Available from edwinhlowepublishing.com, Amazon.com and other book stores.
Printed by Createspace.

Acknowledgements

Henry C. Brown

This book came about because a collector, me, happen to acquire some Soviet Makarov pistols. Using the limited amount of information available and the pistols on hand as study models, I was able to start forming questions that led to a few answers and more questions. This book is the result. Yet this work would never have come into being without the generous help given by numerous individuals. They need to be thanked.

First on the list to be thanked is my family. My lovely wife, Dr. Doris Brown, my daughters Ariel Duhadaway (who singlehandedly rescued this project from disaster), Rhea Votipka and husband Dan, Hana the hand model, and my granddaughter Joanna Duhadaway, the assistant map maker, all provided an incredible amount of support. Other friends and acquaintances I'd like to thank for their contributions are Dr. Cameron White, P.D. Neel, Jay Hart, Matt Cousins, Steve Mullinex, John Moss, Jan Wicker, Alexander Nevarov, Lief Christenson, Andrey Danko, Brian Dressen, Marty Kinnee, Chuck Neely and the folks on GunBoards.Com Makarov forum. Finally, I like to give glory to the author and finisher of my faith, Jesus Christ who listened to a lot of prayer.

Cameron S. White

This book is wholeheartedly dedicated to my father, the late Earl Bruce White, Ph.D. who was my mentor and best friend. His intellectual curiosity, knowledge, and passion for firearms left an indelible and lasting impression on me. His unfailing support and faith in me was the impetus for many of my personal and professional accomplishments, including writing this book.

I would also like to acknowledge the following people who have spurred my interest in Makarovs and related topics and who have helped me grow as a collector through their kindness, encouragement, generosity and knowledge. Some have shared their treasured items with me, others have passed on their knowledge, a few are shooting and gun show comrades; all have helped me appreciate different aspects of the Makarov.

Bob Adams - Albuquerque, New Mexico; Dennis Brandenburg, R. - New London, Ohio; Henry Brown - Roanoke Rapids, North Carolina; Ben Cothran; Terry Edwards - Palmyra, Nebraska; Roger A. Finzel - Albuquerque, New Mexico; T. Jevicky - Home, Pennsylvania; Graham Johnson; Randy T. Kohl, M.D. - Nebraska; Roman James Liekhus - Lincoln, Nebraska; Dieter Marschall – Germany; Moderators of, and contributors to Slim Tim's Makarov Forum on www.gunboards.com; Philip Duane Neel - Oklahoma City, Oklahoma; Alexander Neverov; John Shank – Clymer, Pennsylvania; Richard Thomas, Ph.D. - Lincoln, Nebraska; Ed Tinker.

Finally, a very special acknowledgment and thank you is due to our editor, Edwin H. Lowe, whose meticulous nature, knowledge of firearms, and publishing expertise made this the best work possible. We also appreciate his steadiness, professionalism, and inquisitiveness that were present throughout this project. He consistently went above and beyond and exceeded our expectations from beginning to end.

Contents

Russian Federation soldier firing a Makarov in the standing position prescribed in the instructional manual.[1]

East German Policemen of the Volkspolizei on the range. Note the officer in the foreground with an 'alternative' position of the left hand.[2]

Introduction

On April 12, 1961 Soviet Cosmonaut Yuri Gagarin orbited the Earth.[3] With this feat he made history as the first human in outer space. With him in his Vostok spacecraft, as part of his survival equipment, was a 9x18mm caliber Makarov pistol. Thus the Makarov itself became the first firearm in space. While this certainly was a 'high point' in the long career of this Soviet service pistol it is not the beginning or the end of the Makarov story. The Makarov pistol has seen action all over the world in all climates and terrain.[4] It has been called the "world's most popular handgun in terms of sheer numbers".[5] It has been estimated that in Russia alone, five million Makarov pistols had been produced as of 2002.[6] While that number alone may give pause for thought, if one considers that Russian production of Makarov pistols still continues to this day, and one factors in the untold numbers made in Bulgaria, China, and East Germany, the grand total manufactured to date points to a much higher figure.[7]

Yet, despite these impressive numbers, the Makarov pistol is still somewhat obscure in the West. This lack of familiarity is almost certainly a legacy of the Cold War. Very few trickled into North America during the four decades when that undeclared conflict was at its height. Rare and with an unobtainable cartridge, the Makarov was only a grainy picture in books devoted to the study of small arms.

All this changed with the fall of the Soviet Union in 1991 and the political shake up of Eastern Europe. Batches of Makarov pistols and 9x18mm ammunition were quickly imported into the United States and sold commercially. Its merits began to be appreciated by shooters and a growing interest developed with collectors. Still, a great deal of confusion, misinformation and ignorance remains concerning this classic pistol of the Cold War.

In attempting to counter this, it would be best to start at the beginning. This book is the first of a two volume series of research and collector's observations of the Makarov pistol. Fittingly, this volume starts where the Makarov story began: with the Makarov's conception, design and initial production in the Soviet Union in 1951 and its ongoing production in the Russian Federation, and the first licensed production of the Makarov outside the USSR, in the German Democratic Republic in 1958.

The primary goal of this book is to document the history and variations of the Soviet and East German Makarov pistols in a more thorough manner than has been done previously, from the perspective of the collector. This book was written from the perspective of collectors for the benefit of the collector community. Photographs are often more descriptive than words, and they have been used liberally in this book. The pistols and accessories pictured are from private collections in the United States. The information in this book is based on firsthand observations and research from print references and internet sources, and all sources are fully referenced, both in the interest of scholarship and for the benefit of further research.

The hope is that this compilation of basic information will create a base of information, photographs and reference sources that other collectors and researchers can build upon. The ultimate goal will have been realized if this book is well received and becomes a frequently used reference source.

Henry C. Brown and Cameron S. White
2016

Russian Federation soldier with regulation lanyard and holster with cleaning rod for the Makarov.[8]

9mm Makarov Pistol Model of 1951

9ММ Пистолет Макарова системы 1951 г

Soviet Union & Russian Federation

Henry C. Brown

Enter the Makarov: Designing a Legend

The year was 1945. The Soviet Union and her western allies had finally crushed Nazi Germany and Imperial Japan. For the Soviets especially, the cost in lives, material and national infrastructure had been devastating. Yet, almost before the last shots had been fired, the Red Army announced competitions for a new series of modern small arms based upon the lessons learned during the Great Patriotic War. One of these competitions called for the development of a new universal service pistol.[9]

The *Tulskiy Tokareva* (Тульский Токарева) - Tula Tokarev Model of 1933, or TT-33, had soldiered through the war years with good effect. However, as early as 1938, military and industrial commissaries had called for a replacement. Battlefield reports pointed to specific problems that had surfaced when the TT-33 had been fielded. These included inadvertent magazine release, defects in the trigger, short mainspring life and unsuitability for use with the then popular pistol ports in armoured vehicles. The brutal school of the Eastern Front had taught the Red Army, Soviet Navy and Soviet Air Forces some hard lessons which they intended to apply to a new generation of small arms. Likewise, Soviet industrial workers and engineers had met the severe demands of wartime production and stood ready to apply their considerable abilities to meet new challenges.[10]

The new pistol would be suitable for all branches of service. It could be offered in a 7.62 mm or 9 mm version. The new cartridges fielded in the pistol had to equal the terminal ballistics of the Tokarev's 7.62x25mm round. It was required to be a smaller and lighter handgun than the TT-33, but inherently more accurate. Finally, reliability of the new weapon had to approach perfection in all conditions.[11]

Several engineering teams headed by established arms designers such as N.F. Tokarev (1871-1968) and S.G. Simonov (1894-1986) submitted prototypes for testing. Nicolay Fyodorovich Makarov (1914-1988) submitted prototypes in both 7.65mm Browning and a new 9x18mm cartridge developed by Boris Semin (1911-1982). In 1951, Makarov's 9mm pistol was officially adopted. As was common with most Russian and Soviet small arms, Nicolay Makarov's name became forever tied to his creation and its new cartridge. The new service pistol would be called the *Pistolet Makarova* (Пистолет Макарова) - 'Makarov's Pistol' - or simply, the PM (ПМ).[12]

Late production TT-33, dated 1952.

The Development of the Makarov Pistol (1947-1957)

As with any new design, the Makarov pistol was put through a trial process that included testing and refinement, as well as development of the production techniques required to build it. Typically with Soviet small arms, existing weapons were produced concurrently with newer systems designed to replace them. This would ensure that the new system was performing as required and could be manufactured in quantity and to the standards of quality desired. Such was the case with the pistol the Makarov was destined to replace, the 7.62x25mm TT-33 Tokarev. The official adoption of the Makarov by the Soviet Union was in 1951, yet examples of the Tokarev can be observed with 1952 and 1953 dates.[13]

Tula & Izhevsk, USSR.

The initial production of the Makarov was undertaken in the city of Tula (Тула). Approximately 20 to 30 pistols were manufactured. Photographic evidence shows that these pistols bore the Tula Arsenal mark on the left of the frame and black grips with an entwined "T" and "M" moulded into the top portion.[14]

By 1949, production of the Makarov was moved to the Izhevsk Mechanical Factory Number 622.[15] Known as IZHMECH (ИЖМЕХ), the factory was located in Izhevsk (Ижевск), a city closed to western visitors and located deep in the Ural Mountains, 845 miles east of Tula. Izhevsk had a long established tradition manufacturing rifles and pistols for military and commercial markets. With its extensive production facilities and well trained work force only just recently finished fulfilling the enormous production demands made during the Great Patriotic War, the city was a natural choice as the arsenal for a new generation of small arms. It is noteworthy to mention that the TT-33, SKS 45 and the AK-47 were also being produced simultaneously at Izhevsk in the period from 1949-1954. A batch of roughly 5000 IZHMECH built Makarovs were completed in 1949 for trials. One pistol from the 1949 production batch was engraved as a presentation piece for 'The Boss', Premier of the Soviet Union, Joseph Stalin upon the occasion of his seventieth birthday.[16]

IZHMECH factory marking.

Firing Position from *Manual of Instructions for Use and Maintenance for 9mm Makarov Pistol.*[17]

The large scale trials of the new pistol were deemed successful, and while some changes were directed to be made, the design had proven itself to be sound. In 1953, the Makarov pistol entered full scale production.[18] One obvious change in the 1953 production pistols was the use of brown Bakelite as a new grip material. Makarov pistol grips prior to this had been black with checkered panels on the both sides of the grip and the back strap. Testing of the original trials pistols had resulted in irritation to the firer's hand from the checkering on the back strap. The brown colored grips had a smooth back strap designed to remedy this. Another obvious change occurred to the top portion of the trigger guard. The earlier guns had featured prominent 'wings' on either side of the trigger guard where the take down lug projected into the frame. While a somewhat elegant example of the machinist's work, these 'wings' and the corresponding machined recesses in the frame above the trigger guard were deemed an unnecessary feature and eliminated.[19]

A less obvious feature of the Makarov that was changed in 1953 was the automatic slide release. The special slide release mechanism allowed the slide, when held opened by the slide stop, to close when a loaded magazine was inserted into the pistol. This chambered a round automatically. The automatic loading assembly comprised of three parts. The first was the slide stop which featured a 'J' shaped projection below its mounting pivot. This projection worked in conjunction with a small tab found on the left side of the magazine which engaged it and forced the slide stop down, allowing the slide to go forward as the magazine locked into position (the magazines at this time had the opening on the left side narrowing towards the top, placing the projecting tab directly underneath the slide stop). The third part of the assembly consisted of the grip, which was relieved underneath the left panel to provide clearance for the

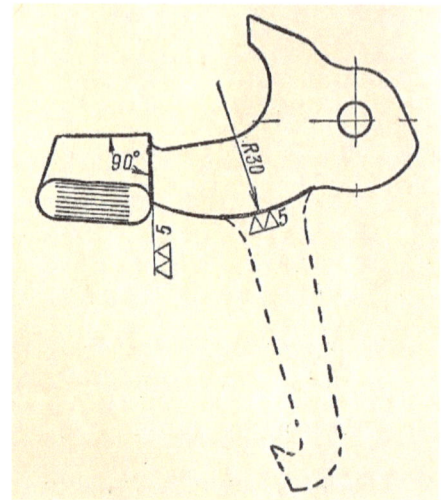

Illustration from *9mm Makarov Pistol (PM) Handbook for Repairs*, depicting original slide release with automatic activation hook.[20]

magazine and slide stop to operate. Problems with accidental firing of the pistol surfaced during the trials, so the assembly was removed from the pistol. A new slide stop was produced without the lower projection, magazines without the tab were designed with a full rectangular opening on the left side, and the interior relief cut in the grip was eliminated. Existing slide stops were modified, and magazines simply had the tab removed. The early grips required no modification.

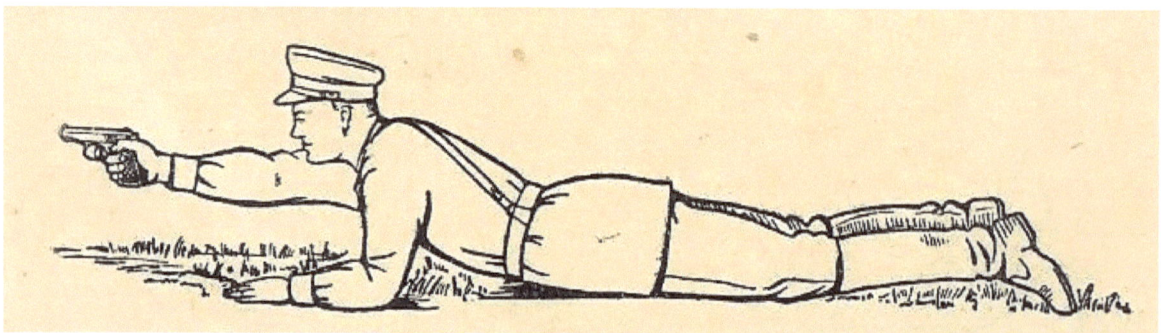

Prone Firing Position from *Manual of Instructions for Use and Maintenance for 9mm Makarov Pistol.*[21]

1953 production Makarov.[22]

1953 production Makarov. Note the protruding machined 'wings' on the top of the trigger guard, and the corresponding recesses on the frame.[23]

Other small but important modifications were also introduced. At this time the mainspring was strengthened by increasing the width of the portion powering the trigger.[24] It was found that retaining the mainspring with the same screw that secured the grip was insufficient. The solution was to cut a notch on either side of the back strap below the threaded screw boss and create a metal clamp to slip up and retain the main spring independently of the screw. The combination sear and slide stop spring was given a projection to fit into a hole in the shank of the sear, rather simply resting on it. The angle of the interface between the trigger bar and hammer was improved.

Mainspring retaining clamp.

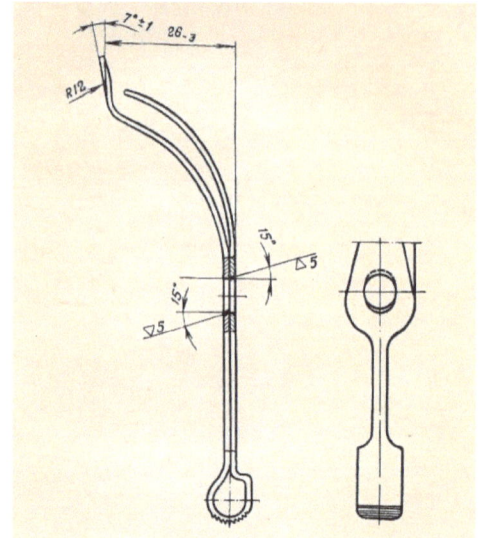

Illustration from the *9mm Makarov Pistol (PM) Handbook for Repairs* of the improved mainspring adopted in 1953.[25]

The pistols produced in 1953 reflected most of these changes incrementally as they were implemented. The 1954 production pistols incorporated all of the new improvements. Often their assembly involved both new made parts and existing parts reworked to the new standard. Production continued in this fashion in order to meet the tremendous requirements of the Soviet armed forces, the police and the numerous internal security services.

Makarov Pistol Model of 1951 Specifications

Caliber:	9x18 mm Makarov (0.363 x 0.71 in)
Barrel:	4 grooves, right hand twist, chrome lined
Barrel length:	93 mm (3.67 in)
Weight with empty magazine:	730 g (25.75 oz)
Weight with full magazine:	810 g (28.57 oz)
Pistol Length:	161 mm (6.34 in)
Magazine capacity:	8 rounds
Finish:	Blue oxide
Grip:	Brown Bakelite
Practical Rate of Fire:	30 rounds per minute

The Makarov Perfected (1955)

The experiences with the early production pistols eventually led to a major revision in the manufacture of the Makarov that married all of the improved components with a newly designed frame.[26] Gone was the flat section of the frame above the trigger that ran parallel with the bottom edge of the slide. With the deletion of the 'wings' on the top of the trigger guard, there was now a simple lug that fitted into the frame to retain the slide. Because of this, the recesses in the frame into which they fitted were no longer necessary. From behind the top of the trigger to the front, the frame now gracefully arched upwards. This greatly simplified the machine work needed to produce the frame with the extra benefits of reducing weight and providing more room in the trigger guard. This new frame was introduced in 1955 and set the standard for all the Makarov pistols to be produced afterward, both in the Soviet Union and elsewhere.

Revised frame shape and simplified trigger guard.

The pistol was of all steel construction with the exception of the reddish brown Bakelite one-piece grip which in itself featured steel reinforcement. Manufacturing was primarily done by machining the components from solid steel forgings. The only parts not made so were the springs formed from round wire, as well as the slide stop, grip screw detent bushing in the grip, and the mainspring which were all of pressed steel. The barrel was given a hard chrome lining for ease of maintenance and longer barrel life. The finish was a deep satin blue oxide applied to all the steel parts exhibiting a surprisingly high standard for a weapon designed to be produce in the millions. The result of all this effort created a very tough, reliable handgun able to shrug off abuse and do the job it was designed to do in a variety of climates.

An interesting feature of the Makarov design is that many parts perform more than one function. The slide stop's top portion also acts as the ejector. The mainspring powers the hammer, trigger, and is also the magazine catch. The trigger guard is used as a disassembly latch. These features, combined with the unitizing of other parts into sub-assemblies, results in a firearm of remarkably few parts. With the exception of the barrel and trigger guard assemblies, the Makarov may be stripped down completely with little more than the magazine base as a makeshift take down tool, though the issued combination tool/cleaning rod makes this task easier. Virtually all the parts have sufficiently consistent tolerances to be considered 'drop in' replacements. Parts manufactured during Soviet or later Russian Federation production of the military issue Makarov, as well as licensed production parts manufactured in the People's Republic of China, the German Democratic Republic and Bulgaria will normally interchange.

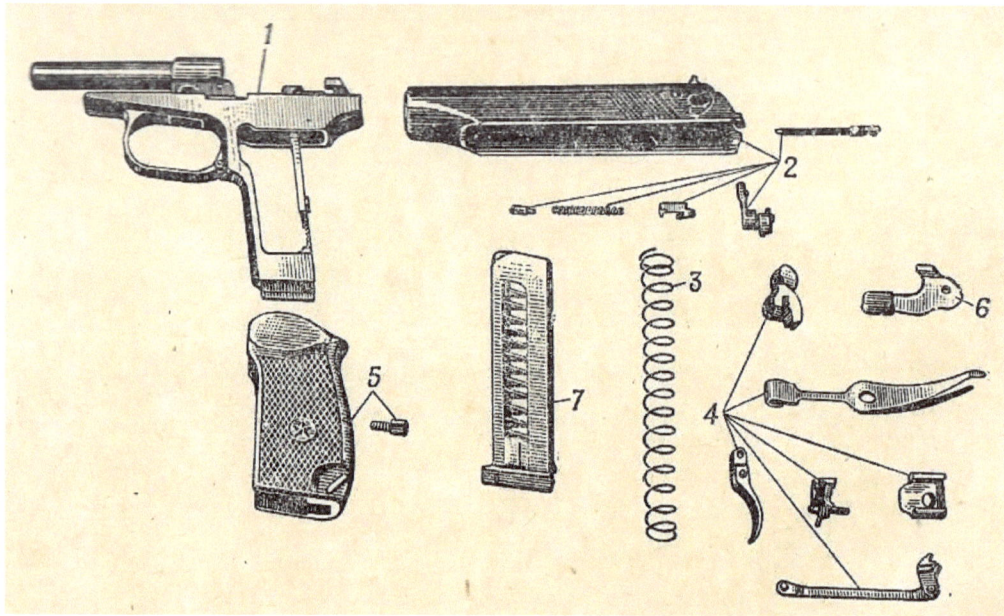

Exploded view of Makarov from *Manual of Instructions for Use and Maintenance for 9mm Makarov Pistol*.[27]

1. Frame and Barrel
2. Slide, Firing pin, Safety, and Extractor with its spring and detent
3. Recoil Spring
4. Hammer, Mainspring, Clamp, Sear, Trigger, and Trigger Bar
5. Grip and Screw
6. Slide Stop
7. Magazine

Serial Numbers and Markings

The Makarov's serial number consisted of a two letter Cyrillic prefix followed by three to four numerals. These formed production blocks of pistols marked - in theory - up to 9999 at which point the alphabetical prefix would change and the numbers would cycle through the new production block. The expression "in theory" is used here because serial numbers containing one or two digits have not been observed by collectors. Each year's production saw several different Cyrillic prefix combinations used, and prefixes could be reused in later production pistols, so technically their date of manufacture needed to be included as part of the serial number.[28] The frame was machine engraved with the serial number, arsenal mark and date on the left side above the grip. The left side of the frame featured the engraved serial number, while the safety displayed the last three digits. Because the process of engraving the serial numbers, arsenal mark and date involved actually cutting into the surface of the frame, slide and safety with a rotary milling tool, it had to be done 'in the white' prior to any heat treatment hardening of these parts. It is common to see the last two or three digits of the serial number scratched or electro-pencilled by hand upon the surface of the various internal parts as well.

'P' (1963) dated Makarov with original magazines serial numbered to match.

The year that any individual Makarov was produced during the first decade of manufacture was indicated either by an actual four digit date, such as '1960', or a Cyrillic letter date, eg 'M'. There has been speculation that using the letters to date pistols would create confusion in western military circles regarding the time frame the Soviet Union was producing them. If, indeed, such was the case, the ruse succeeded admirably. There is still considerable confusion among collectors to this day over the correlation between letter and date.[29] To add to this confusion, in some years, both numerically and letter dated examples were manufactured. Yet, while this makes assigning a specific year to any given letter a frustrating experience, the reality is that the time frame involved was small. The system of using letters to date Makarovs only lasted roughly from 1956 to 1963. From 1964, the numeric dating system became the norm.

Cyrillic Letter Date Codes [30]

Д	1956	И	1957	К	1958
Л	1959	М	1960	Р	1963

In 1962, the letter dating was dropped in favor of numeric dates, but for 1963, the letter date 'P' as used for the last time. Note the electro-pencilled serial number on the magazine (top) and the final two digits of the serial number electro-pencilled on the sear (bottom).

A wide range of other markings are to be found on Soviet production Makarovs. These can be found on the back strap of the frame and the sides of the frame located under the grip, the top front of the trigger guard, on or near the barrel breech, and the underside of the slide. These cryptic marks normally consist of numbers or letters, often with in circular, square or diamond shaped outlines. Most tend to be very lightly struck and consequently are difficult to read. Unfortunately, interpreting the meaning of any of these symbols is, at best, a guess. Similar stamps can be found on other Soviet made small arms of the same period and are most likely inspection and quality control marks.

A less obvious change occurred in 1960, when the cut out in the rear of the frame was eliminated. Note the 1961 dated frame (top) and the 1959 dated frame (bottom). Note also the inspection markings on the back straps of the pistols.

Л (1959) dated Makarov with various manufacturing marks.

This is not to say that an educated guess cannot be stated as a certainty in some instances. The face of the rear sight was normally marked with a number, and '3', '4' or '5' are the most commonly found, though higher and lower numbers exist. These numbers indicate the height of the sight. The point of impact of the fired bullet can be raised by installing a higher numbered sight or lowered by installing a lower numbered sight. German Democratic Republic armorer's spare parts kits contained a range of numbered rear sights up to '8'.

Rear sight with the number '3' denoting the height of the sight.

Other numerous observations leads the author to conclude that the definitive proof marks on Soviet Makarovs, while not verified by any Russian source, are almost certainly a stylized Cyrillic 'TK' in a circle and a 'ПС' (PS) within a circle. The 'TK' was apparently an earlier proof mark used prior to 1960. After that time the 'ПС' marking is found. Initially both were stamped under the slide, on the left side of the frame lug which houses the barrel. In 1965 the 'ПС' was moved to the frame just after the date of manufacture. Again, many are faintly stamped and only partially legible. After 1959, the 'ПС' is the one marking consistently found on Soviet Makarov pistols.

It was not a standard Soviet military practice to mark small arms with unit designations or any other markings other than the serial number. However one example of a 1986 dated Soviet Makarov has been observed with the number '32' painted in white on the frame immediately below the hammer. The '32' is painted upside down such that it would only be legible if the pistol were resting on the top of the slide. This is most likely an expedient way of quickly identifying a unit's weapons for issue from a storage rack. Weapons so issued would usually include two magazines and sixteen cartridges, a holster and cleaning rod.[31]

Makarov dated 1986 with painted rack number on frame.

One further group of markings will occasionally be found on Soviet Makarovs. These are associated with arsenal refurbishment of the pistol.[32] On the left of the frame, in the general area of the slide release lever, a square box with a diagonal line will occasionally be found. This marking is often found on numerous other refurbished Soviet small arms. As expected, this stamp has been found on earlier production Makarovs. In at least one case, an additional marking has been applied below it. It has been observed that a box with a marking inside may also appear on the right frame directly matching the position of the refurbishment stamp on the left side. There one example, a letter 'M' dated pistol, that has the marking on the right side of the frame, but no corresponding mark on the left. As with many of the stamped markings found on Soviet Makarovs, the impressions are often struck in such a manner as to be indecipherable.

A pair of 1963 dated Makarovs. Note the arsenal refurbishment mark on the top pistol, forward of the magazine release on the left side of the frame (top). There is a corresponding arsenal refurbishment mark on the right side of the frame on the same pistol (bottom, left).

Ongoing Production Changes and Modifications (1965-1990)

By the time the relatively insignificant relocation of the proof mark was made, Makarov production was in full swing. Enough pistols had been manufactured to apparently satisfy the requirements of the armed forces. This is suggested by the appearance of Soviet manufactured Makarovs around the world during the 1960s as the Soviet Union provided weapons and technical support to various client states. In the case of the German Democratic Republic, this involved a license for full production. For other countries in Eastern Europe, the Makarov was either adopted outright or a native design in 9x18mm caliber was developed as part of the Warsaw Pact standardization program.[33] Further afield, numerous Soviet made small arms, old and new, began to appear in regional insurgencies and conflicts throughout the world as the Soviet Union sought to extend its influence and control. Along with the SKS, AK-47 and the RPG, the Makarov pistol made its debut on the world stage in conflicts in Africa, South East Asia, and the Middle East. Though never seen in the sheer numbers of the AK-47, the Makarov became a symbol of prestige and authority for anyone considered important enough to be issued one.

It is perhaps because of this widespread use of the Makarov in jungle, desert, and veldt as well as within the Soviet military itself that a number of relatively minor changes were made to the pistol during its production life up to the demise of the Soviet Union. Some of these alterations were to improve the function of the pistol, making it even more reliable no matter what part of the world it was to be fielded. Other changes were obvious attempts to both simplify production and reduce manufacturing costs.

Beginning as early as 1967, a significant change appeared in the manner with which the serial numbers, arsenal marks and dates were applied to the frame and slide. Prior to this time, these markings were machine engraved, which was a time consuming process requiring special tooling. Additionally, it had to be done prior to any heat treatment and final finishing, which meant wasted effort on damaged or rejected parts. The solution to this was the 'dot peening' process.[34] Also known as the 'micro percussion method', this involves a machine with one or more hardened styluses rapidly striking a surface to create groups of small dimples or 'dots' in the shape of alphanumeric characters. While not as deep or sharply defined as engraved markings, the dot peening marking process was quick, legible, and could be applied to a finished pistol. As with most modifications made to the Makarov, the transition to the new marking method took place over a two or three year period as pistols already in production were completed and older parts were used.

Along with the adoption of this new marking system, the serial numbers engraved on the safety were eliminated. Photographic evidence points to this occurring roughly during the same period. Some pistols dated 1968 have serial numbered safeties and some do not. By the time of 1969 dated pistols, serial numbers are no longer evident on the exterior of the safety lever. Safety levers on later dated Makarovs can be found with the serial number of the pistol scratched on the interior surface. Aside from the new method of marking the slide and frame, another obvious change occurred in the shape of the depression, or detents, in which the safety spring snapped into on the left side of the slide as it went from 'fire' to 'safe' and back again. Until 1973-1974 these had been round but changed to an oval after this time. Sometime around this period, a hole in the raised portion of the safety lever appeared in production. An existing blind hole on the underside of the safety that seated the front of the detent spring was now drilled completely through the thumb lever.

Makarov dated 1974, showing the first year of production using the new safety detents

In the 1973-1974 period another less obvious modification appeared, apparently to improve the reliability of the pistol. This was the enlargement of the ejection port on the slide. The front portion of the opening was expanded slightly towards the muzzle, and the upper part of the opening was raised by 1 mm. This necessitated the reduction of the original 3.5 mm width of the sighting rib on the top of the slide by 1 mm. From this period onward, the sighting rib on the top of the slide was 2.5mm wide. Machine work on the interior of the slide was simplified. A comparison of the front of the slide of earlier production Makarov pistols and those produced after 1973-1974 will show an increase in size of the small flat portion of the slide directly underneath the muzzle. This is sometimes referred to as a 'beard'. This profile from the end of the frame up to the muzzle, is a consequence of less metal being removed from the bottom of the slide. The size of the 'beard' became larger over time as less steel was machined off during production.

Beginning in about 1976, a new innovation in production techniques appeared. Hammers, safeties, and triggers started to be produced by the investment casting process. These new parts may be readily identified in the case of the hammer and safety by the deep blue/black color of the investment cast parts, in contrast to the red tinted finish found on milled parts. A closer comparison of milled and cast parts reveals sprue marks left over from the casting process, as well as the safety having a more rounded lever and vertical serrations. All parts milled or cast are completely interchangeable. Pistols examined bearing dates from 1976 to 1983 contain a mixture of milled and cast parts, with the later gradually coming to predominate.

Makarov dated 1976 in original shipping box with a brown
waxed paper wrapping and hand written serial number.

Comparison of Early and Late Production Parts

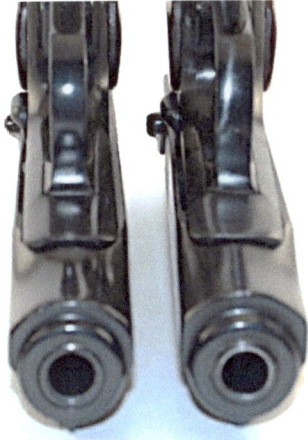

'Beard' profile 1961 (left) & 1987 (right).

Soviet era milled sear & post-Soviet era investment cast sear.

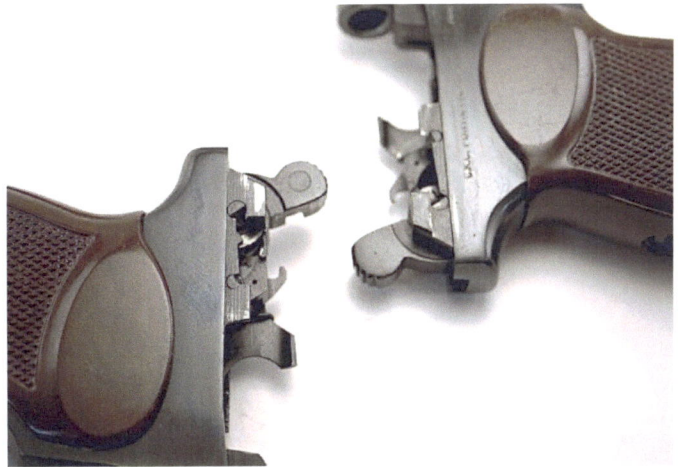

Slide releases: 1953-1982, 1983-1990, 1990-present (left to right). Note the change to the shape of the ejector arm portion of the slide stop from 1983.

Investment cast hammer 1976-present (left), milled hammer 1951-1976 (right).

Safety levers: 1955-1972, safety with hole 1973-1982, cast safety 1983-present.

Early production 3.5 mm sighting rib (left). Later 2.5 mm sighting rib necessitated by the enlarged ejection port (right).

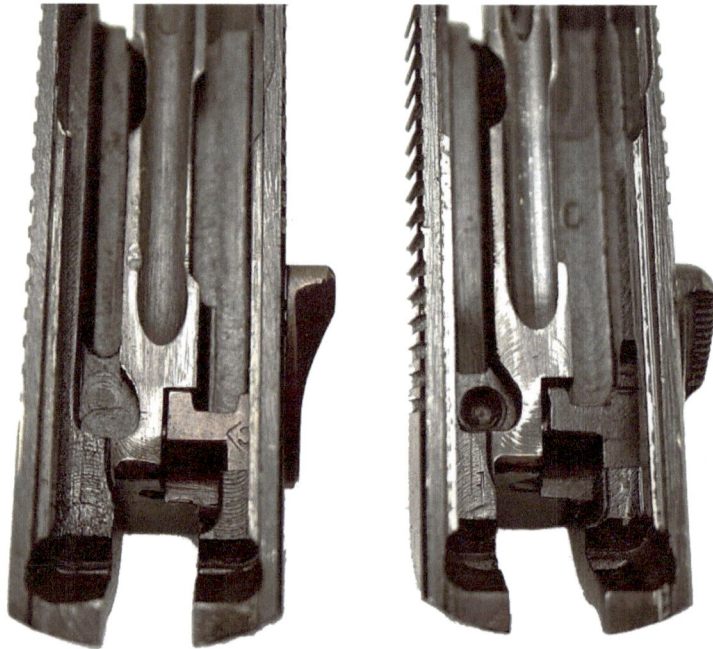

Earlier production slide (left). 1974 slide with simplified machining (right).

At approximately this time, though it must be emphasized that no clear time line has been established, a new pattern of magazine appeared. The original magazine with its flat top 'hump' or tunnel which allowed clearance for the slide stop actuating tab on the left side of the magazine was replaced. The new pattern featured a simple slit in its place.

The earlier method of stamping the magazine spine with the serial number, ended perhaps as early as 1967, after which serial numbers were written with an electro-pencil on the left side of the magazine parallel to the rear face. Older magazines can be found with electro-pencil serial numbers and reflect, in all likelihood, new old stock being assigned as replacements to a pistol.

Often older magazines which have gone through one or more refurbishments will bear stamped numbers and then one or more other serial numbers applied as they were reissued. While the original stamped serial number was placed on the spine of the magazine parallel with the base, the electro-pencil marked serial numbers were applied parallel to the length of the magazine. Normally this was done on the left side of the magazine, though at least one example has been noted with duplicate serial numbers applied to the spine, as well as the side.

The new style magazines bear few markings, if any at all. On some, a cryptic oval marking appears on the left side above the base. Serial numbers will often be present scratched or electro-pencilled onto the left side. Bulgarian production pistols made beginning in 1976 have identical magazines. This makes any attempt to classify one of the late magazines Russian or Bulgarian a lesson in frustration.

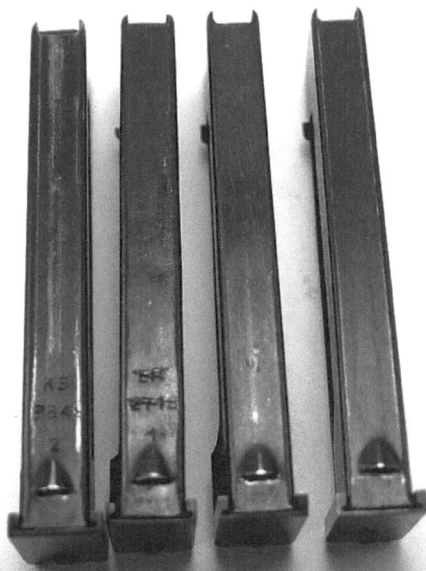

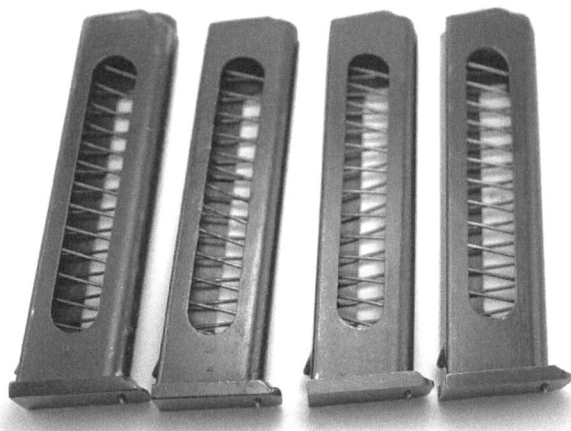

Magazines: Early magazine with original serial number; early magazine with arsenal refinish, re-serial numbered; early magazine bearing electro-pencil serial numbers; new pattern magazine introduced approximately 1976 (left to right).

Machine cuts in the frame surrounding the trigger guard lug well (left). 1984 or later frame without machine cuts (right).

Makarov dated 1987 with post-Soviet era Russian commercial holster. Note the two digit date and increased number of slide serrations.

The final years of Makarov production by the Soviet Union witnessed a small number of other new developments. In 1983 a new shaped ejector arm was adopted as part of the slide stop. This section is noticeably thinner in profile than the earlier production ejector arm. This was followed in 1984 by the elimination of two rounded machined cuts in the interior of the frame above the trigger guard lug well. A more obvious alteration was the elimination of the first two digits in the pistol production year, for example, a 1985 production pistol was marked '85'.

The number of grip serrations cut into the slide varied over time. From 1955-1983, the standard pattern was ten serrations on the left rear of the slide and seventeen serrations on the right rear of the slide to create a secure grip to retract the slide. This pattern was standard for the next 27 years of production. However on some slides made in 1983, there were ten serrations on the left and eighteen on the right. Slides made in 1984 featured seventeen serrations on the left and ten on the right of the slide. Evidence suggests that this number of slide serrations was retained into the middle 1990s when the original pattern was re-adopted.

A final two observations may be made concerning the Makarov pistols made in the waning days of the Soviet Union. The numerous inspector and quality control markings found on pistols manufactured in the early years declined over time. By the late 1980s they were often reduced to one or two. Secondly, the quality of the exterior of the finish, while by no means poor, evolved into a less glossy sheen. None of this hampered the performance of the Makarov. It remained a well-engineered design with solid dependability.

In 1991 the Union of Soviet Socialist Republics ceased to exist. The government which N.F. Makarov had designed his pistol to serve was suddenly gone. However, this was not to be the final chapter in the history the Makarov. Its service life would extend beyond the existence of the country that had built it, the Union of Soviet Socialist Republics. Newly produced pistols were to be made in an increasing number of variations for the Russian Federation well into the next century. In a final irony, the mysterious Makarov pistol would become available in the West, particularly the United States. There, a new generation of collectors and shooting enthusiasts would soon learn to appreciate the qualities of this "commie gun" from the other side of the Cold War.

The Russian Federation Makarov (1991-2016)

On December 26 1991, the Union of Soviet Socialist Republics ceased to exist. This act reflected the reality of the dissolution of the Soviet Union over the past year as the member republics fell away from their long association. A period of great social, political and economic upheaval began as each of the fifteen former republics sought to follow their own course. While much of this is beyond the scope of this book, it is safe to say that the impact of this turmoil on the huge complex of weapons manufacturing factories in Izhevsk was historic. Izhevsk Mechanical Factory 622, producer of the Makarov pistol, was faced with a seismic change.

In a short space of time, the strict oversight that the Soviet government had exercised since the beginning of the Makarov's production disappeared. Along with it went the economic support provided by the state and the biggest customer for the factory's output. On a darker note, the control and security previously exercised by the Soviet government relaxed as well, giving rise to the potential for theft and abuse.[35] IZHMECH quickly had to come to terms with these issues and learn to survive in a free market economy or close the doors.

Russian Federation soldier with Makarov pistol.[36]

How IZHMECH survived is a somewhat convoluted story. Suffice it to say that it did so, though not without struggles and reorganizations.[37] Some government support still existed along with private investment capital that allowed the factory doors to remain open. The Russian military still needed small arms research and production facilities, but this was not enough. The new emphasis was towards a more self-sustaining operation. The most obvious means towards self-sufficiency was to offer a range of products with an appeal to commercial and civilian markets beyond the borders of the new Russian Federation. Convergent with this was the need to update the Makarov pistol to meet the needs of military, security, and police agencies.

PMM Modernized Makarov

In addition to the numerous modifications made during the Makarov's forty odd years of production with an eye towards simplifying construction and decreasing costs, there was another more radical departure from the traditional manufacturing process. In the mid-1960s the experimental TKB-023 polymer frame Makarov was developed, successfully tested, but ultimately passed over due to longevity concerns.[38] This predated the rise of polymer framed handguns such as the Glock in Western Europe and the United States by some years, but the design was never revived for commercial sale. The real problem was that no matter how many refinements were made to the Makarov, they could not overcome the pistol's greatest deficiency - the design was showing its age.

By the closing years of the 1980s, it was painfully obvious that the Makarov lacked the firepower required for the modern battlefield. The combination of the 9x18mm cartridge and the simple, reliable, inertia or 'blowback' method of operation that had made the Makarov so successful had become a liability. The pistol, when adopted in 1951, was already operating at the practical limits of cartridge power for its design. Three decades later, military and police organizations wanted handguns with the ability to defeat the more effective body armor coming into widespread use. Additionally, the Makarov's archaic eight round magazine capacity seemed paltry beside a new crop of 9x19mm pistols that regularly sported magazines with twice that capacity.

PMM Modernised Makarov.[39]

In 1990, during the waning days of the Soviet Union, an updated Makarov pistol appeared. Dubbed the *Pistolet Makarova Modernizirovannyy* (Пистолет Макарова Модернизированный) - Modernized Makarov Pistol, or PMM (ПММ), it featured a redesigned grip and a heavier frame and slide. More to the point, the frame was wider in order to fit a 10 or 12 round capacity magazine. To load the magazine, a new 9x18mm cartridge was also developed, the 57-H-181CM, which generated approximately 25% more muzzle velocity.[40] To help the simple inertia system cope with the more violent recoil, a series of spiral flutes was cut into the chamber walls. These flutes provided a gripping surface for the cartridge, slowing its rearward velocity to some extent. Ultimately these flutes were deleted, as was the new, more powerful cartridge. There was a well based fear that the more powerful cartridge would be inadvertently fired in a standard Makarov pistol and damage both pistol and shooter. A limited number of PMMs were purchased by the Russian Federation for issue to specialist military and police units. These fire standard pressure 9x18mm cartridges.

In 2003 the 9x19mm Yarygin Pistol or PYa (ПЯ) was officially adopted by the Russian Federation's armed forces to replace the Makarov. Yet, Nicolay Makarov's pistol did not suddenly vanish. It will take many years for the transition to the Yarygin to be completed. The Makarov is still very much a presence in second tier military units and police throughout Russia. It has regularly made appearances in photographs and news casts documenting conflicts in Easter Europe, Africa, and the Middle East.

Baikal IJ70 Makarov

Aside from its use by military, police, and security organizations, there are also numerous versions of the Makarov pistol sold commercially. For most collectors and firearms enthusiasts, one of the first things they notice after unboxing any type of Makarov handgun is the 'Baikal Corp.' marking on the slide.[41] The Baikal Corporation was founded in 1962 as an export group to handle the sale of numerous manufactured goods made in Izhevsk. The products sold under the Baikal brand include a large array of sporting firearms, car parts, investment castings, medical items, and industrial equipment. For the Makarov pistol owner in the United States, the Baikal marking will be found on the various models which were allowed to be legally imported briefly from 1994 to 1996.

The basic model 'Baikal' Makarov was the IJ70. This had an eight round capacity frame with the features found on late production Soviet Makarovs as well as some new features. The 'A' in the model designation stood for the large adjustable rear sight which replaced the simple notched rear sight. The grips were black with the 'Baikal' logo on the left side, and 'Made in Russia' on the right. Occasionally the triangular black plastic PMM style grip was used. The slide release lever with its rounded cross section was replaced with a simple bent tab. Finally, the factory markings were no longer done by the dot peening process, but rather by laser engraving.[42]

Baikal IJ70-17-A

A more advanced model of the IJ70 was also imported. It utilized the wider PMM frame and magazines with a ten or twelve round capacity. The large black triangular PMM grips were used exclusively on this model. The high capacity framed IJ70 grip and magazines were not compatible with the narrower original model, but all other parts were interchangeable. Interestingly enough, the earlier eight cartridge capacity magazines will function reliably in the wider frame.

The markings that denoted the model numbers on imported Makarovs were neither consistent, nor were they consistently applied. One batch of pistols may be marked 'IJ70', another 'PM/IJ70', and some were marked in Cyrillic. On some, the model number may be found on either the left or right side of the frame. Another quirk was found with the serial numbers. Often the original Cyrillic alphanumeric serial number was over stamped with an import serial number in a Latin alphanumeric format. Yet, other pistols exist with the original Cyrillic serial number intact. There are no particular markings on any of the IJ70 variety of pistols that allow precise dating. The month and year of manufacture is only recorded in the original factory owner's manual.

Investment cast frame, typical of PMMs. Note opening below barrel exposing the trigger guard spring.[43]

Baikal IJ70M Acceptance Certificate with date.

List of IJ70 Models

Model	Caliber	Magazine	Sights	Finish
IJ70	9x18mm	8 round	adjustable	blued
IJ70-17-A	9x17mm/.380 ACP	8 round	adjustable	blued
IJ70-17-AS	9x17mm/.380 ACP	8 round	adjustable	satin chrome
IJ70-17AH	9x17mm/.380 ACP	10/12 round	adjustable	blued
IJ70-18A	9x18mm	8 round	adjustable	blued
IJ70-18AS	9x18mm	8 round	adjustable	satin chrome
IJ70-18AH	9x18mm	10/12 round	adjustable	blued

Codes
17   9x17mm/.380 ACP/9mm Kurz
18   9x18mm
A   adjustable rear sight
H   wide frame for high capacity magazines
S   satin chrome finish

It must be stated that the level of fit and finish exhibited on the IJ70 series was not up to the same standards as the Soviet produced Makarov. The polish was not carried to as fine a finish and some parts exhibit roughness of finish and file marks. Occasionally the fit of the safety was not as tight. During the Soviet era, inspectors were in place to ensure tight quality control during all aspects of the Makarov's production. Poor workmanship, use of inferior materials, or deviation from specifications could be quite literally construed as a crime against the state.[44] This all changed with the need to generate a profit in the post-Soviet economy and it ought to come as no surprise that cutting the bottom line affected the production line as well. However, the design was a solid performer made of good materials and worked just as well as the Soviet made predecessors. On April 3, 1996, President William Clinton signed The Voluntary Restraint Agreement. This ended the importation of the IJ70 series of Russian Makarovs into the United States.[45]

Baikal 654K Pneumatic Makarov

During the period that the IJ70 pistols were being imported, another and very different Makarov pistol appeared. This was the Baikal 654K, or pneumatic Makarov. It was not a firearm but a 4.5mm pistol that utilized standard 12 g carbon dioxide cartridges as a source of compressed gas to fire a BB shot. The wider PMM frame was used with appropriate modifications to the frame, slide, hammer, and barrel. The heart of the mechanism was the 'magazine' which was an investment casting which held the CO_2 cartridge, 16 BBs and the hammer activated valve that released the CO_2 propellant. European American Armory Corporation of Cocoa, Florida imported these from roughly 1998 to 2001. Nicknamed the 'Air Mack', this pistol had an appeal for collectors and shooters who wanted a CO_2 pistol as solidly constructed as a real firearm.

Baikal 654K pneumatic Makarov, an early import by EAA.[46]

The Baikal 654K would never be mistaken for a precision target arm, but it came from the same assembly lines that made the centerfire Makarov pistols and shared many of the same parts. This very same rugged construction with the same ordnance grade materials as an actual Makarov pistol, is what doomed its continued importation. In 2001 the Bureau of Alcohol, Tobacco and Firearms of the United States Department of Justice ruled the design could be too readily converted to fire live ammunition.[47] Importation was ordered to cease. Since that time, the design has been reworked in Izhevsk to prevent any such conversion and these newer generations of the Baikal 654K have achieved a worldwide popularity elsewhere. While the occasional 654K has been brought into the United States by collectors over the years, the commercial importation of these interesting air guns has never again been undertaken.

Baikal 654K with narrow frame and magazine, manufactured 2013. Note the recycled military pistol parts.

The first two digits of the serial number of a Baikal 654K correspond to the last two digits of the date. eg '13' for 2013.

Kalashnikov Concern

The current manufacturer of the Makarov pistol is Kalashnikov Concern, a joint stock holding company with a majority of shares, 51%, held by the Russian Federation. Kalashnikov Concern reports that it is responsible for 95% of all Russian small arms production sold both locally and to 27 other nations under three separate brand names. The first is the 'Kalashnikov' brand of military, police and civilian small arms. Next is the familiar 'Baikal' brand of hunting and target guns. Finally, there is the 'Izmash' brand sporting rifles.

Kalashnikov Concern lists several models of the Makarov pistol available to military and law enforcement.[48]

Model	Caliber	Magazine	Sights
PM	9x18mm	8 round	fixed
PMM	9x18mm	12 round	fixed
MP-71	9x17mm/.380 ACP	8 round	fixed
MP-471	10x23mm Traumatic	8 round	fixed

All of these models will be familiar to the collector with the exception of the MP-471. Also called a *Makarych* or 'traumatic pistol', it represents a class of civilian self-defense arms sold in the Russian Federation and elsewhere that fire 10x28mm blank, tear gas, or Traumatic cartridges. The Traumatic cartridge consists of a slightly oblong 10mm rubber ball capable of inflicting a painful but normally non-lethal impact. Traumatic cartridges in 9mm and 12.7mm also exist for *Makarych* pistols. Other versions of the Makarov pistols are sold under the Baikal brand and include the 654K CO_2, the IZH-79-8, and 9mm traumatic pistols. Somewhat confusingly, while the 654K air pistol has kept to a single model designation, many of the current models of the Makarov pistol have appeared marked with differing model numbers over the years. All the more interesting is the possibility that some of these models are actually factory reworks of original military pistols or utilize older components. The older Bakelite grip material has been replaced by a red or black thermoplastic. For the sake of clarity this discussion will limit itself to using current model designations given by the current manufacturer, Kalashnikov Concern.

One further note of interest ought to be mentioned regarding the current production of Makarov pistols. It has been reported that the firms of Berkut & SOBR, and Forte located in Ukraine have been manufacturing Makarovs firing the various blank, tear gas and traumatic cartridges. These are likely reworks of Makarov pistols for local sales.

Kalashnikov Concern warrants new 9x18mm Makarov pistols for 4000 rounds of live ammunition. History has shown that the Makarov pistol, like its stable mate, the AK-47, has a much, much longer service life than this, given proper care and the occasional change of the recoil spring. To anyone familiar with the Makarov, such a long service life in military and police forces should come as no surprise. The pistol exhibits such ease of handling, dependability and rugged construction that it is now endearing itself to a new generation of private owners. What has been surprising is the design's adaptability to new purposes and markets. It is these attributes, along with the massive numbers manufactured, that will ensure that the familiar silhouette of the Makarov will continue to be seen around the world for years to come.

MP-71 Makarov in 9x17mm/.380 ACP, manufactured 2013.

PM Makarov in 9x18mm, manufactured 2013. Note the 'Baikal-442', 'Made in Russia' and two digit '13' date code markings. The grips are made of plastic.

Holsters

Leather Makarov holster.

The classic military holster for the Soviet Makarov is a fully enclosed, flapped holster.[49] Navy and Naval Infantry holsters are black in color. All other duty holsters carried by the military and police are brown. Special holsters for dress occasion will be found made of smooth white vinyl with no spare magazine pouch.

Very early holsters had a body made of canvas with a brown vinyl textured coating and applied leather details. Some examples were coated on both sides of the base fabric and had a separate piece of fabric included bearing factory markings. Others had factory markings stamped directly onto the underside of the flap retaining strap. Later holsters (with exception of the white parade holsters) were manufactured completely of leather. Normally the body of the holster was sewn to a separate piece comprising the flap, but examples will surface that have the body and flap made from a single piece of leather. Also sewn to the body are the belt straps, spare magazine pouch, two loops to hold the cleaning rod, the wide gusset for the butt of the pistol, a strap to draw out the pistol from the depths of the holster and a strap that fastens to a brass stud to close the holster. The interior of the holster had a thin leather seam liner and toe plug glued in. Dates and manufacturer are stamped in ink or pressed into the underside of the flap along with inspection marks. Even with light wear such markings can become illegible.

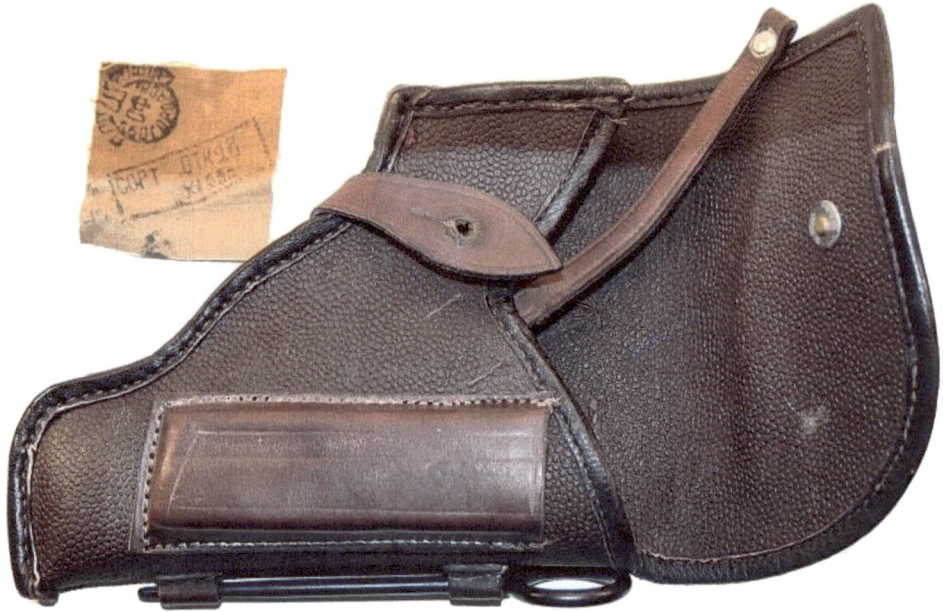

Early pebble finish, vinyl covered fabric body with applied leather details dated 1953 (found on fabric insert). Bogordsk Garment Factory, Nizny Novgorod Oblast, USSR. This factory no longer exists.

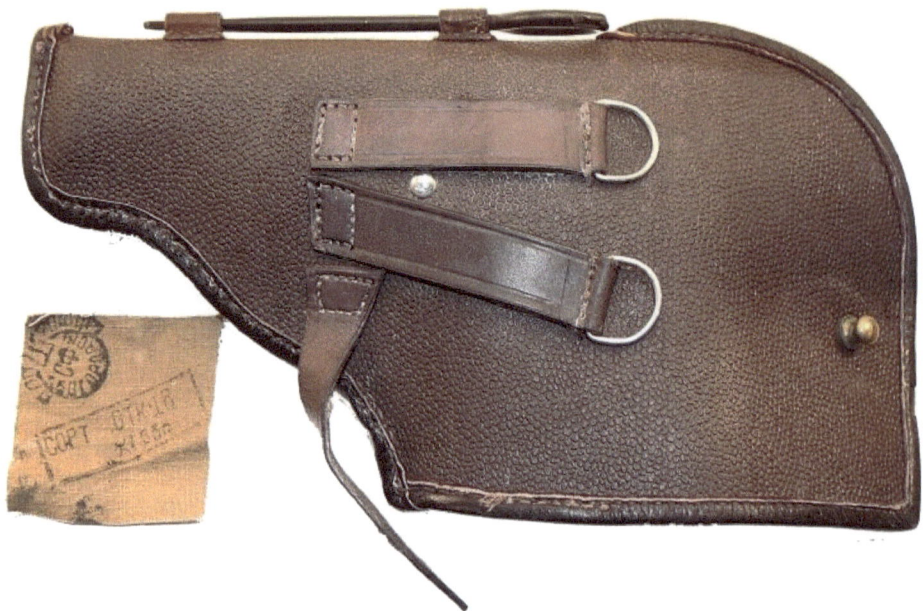

Typical leather construction holster with two piece construction (flap and body), two belt loops and fully stitched. Inside a thick leather toe plug was glued in place along with a thin leather lining that covered the stitching.

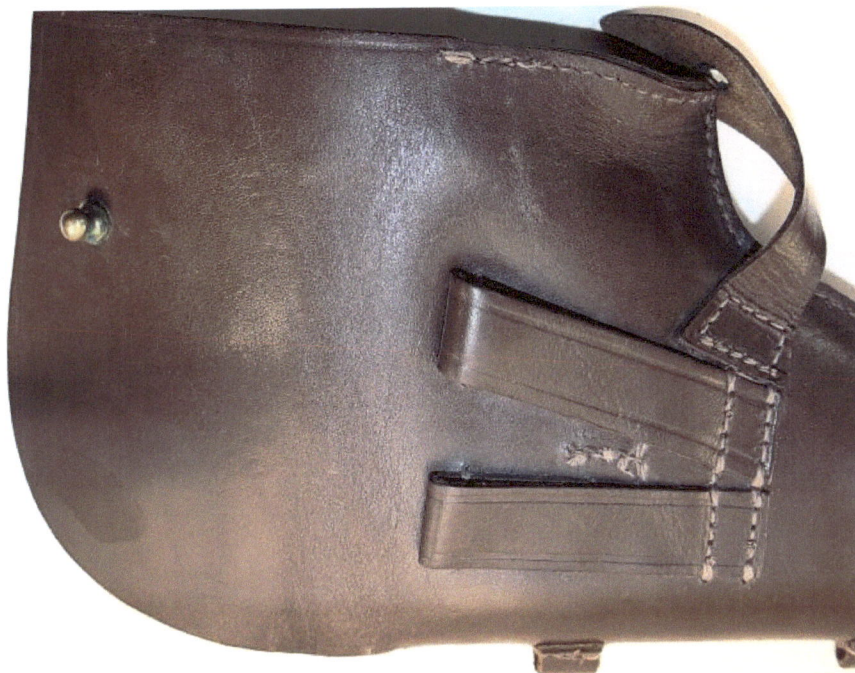

Leather holster dated 1971, with one piece leather construction body and flap.

Rivets were introduced into holster construction in 1978.

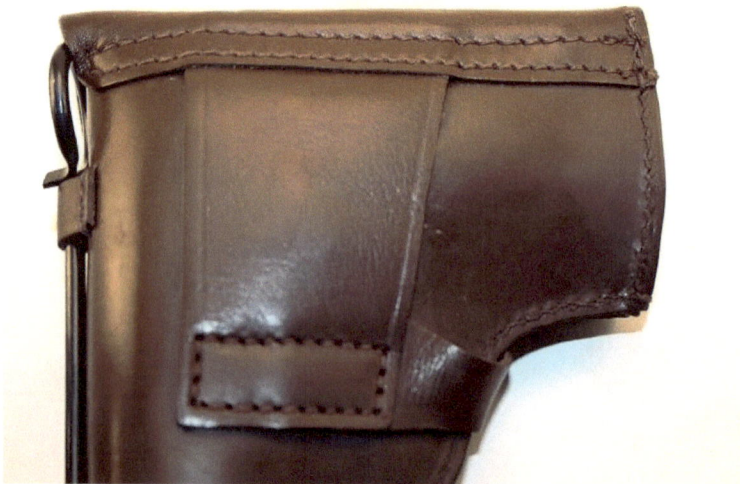

The final version of the Makarov holster with
simplified construction appeared in 1983.

Leather Makarov holsters were originally of stitched construction. From 1978, holsters began featuring chrome plated rivets on some components, a feature which continued with the new simplified holster beginning in 1983. The newer holster had one large belt sewn to the holster with the same stitches used to secure the flap. Similarly, the flap retaining strap was now stitched together with the bottom of the belt loop. The following years saw the quality and thickness of the leather used to manufacture holsters decline, and some samples. appear to be made of a leather and synthetic laminate.

Typical ink stamp located on the inside of the flap,
dated 1976, with the 'OTK 10' inspection mark.

The OTK inspection mark is an abbreviation of *Otdel Tekhnicheski Kontrolya* (Отдел Технически Контрол) or 'Department of Technical Control'. It is followed by a number which represents an individual inspector, and a two digit date.

Holster Variations

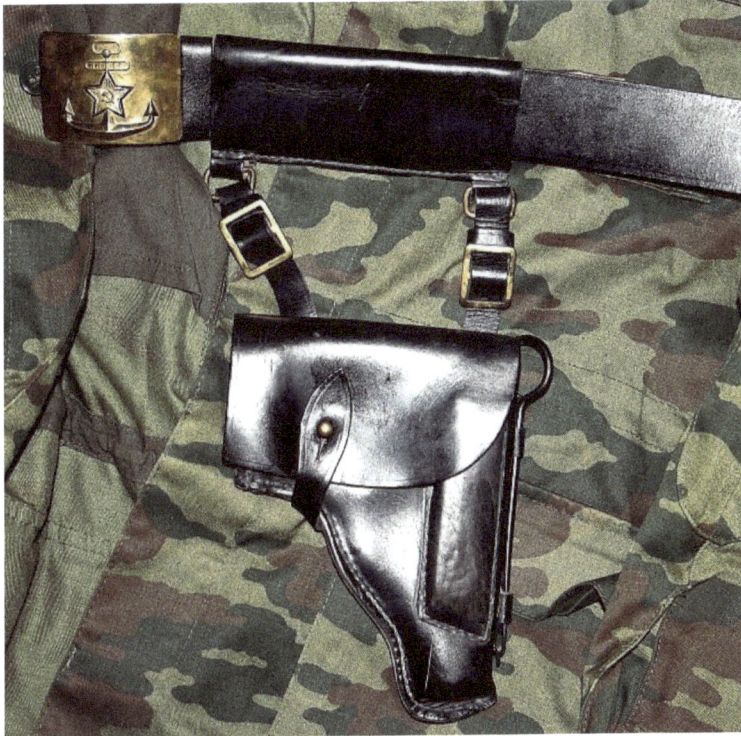

Black Soviet Naval Infantry holster.

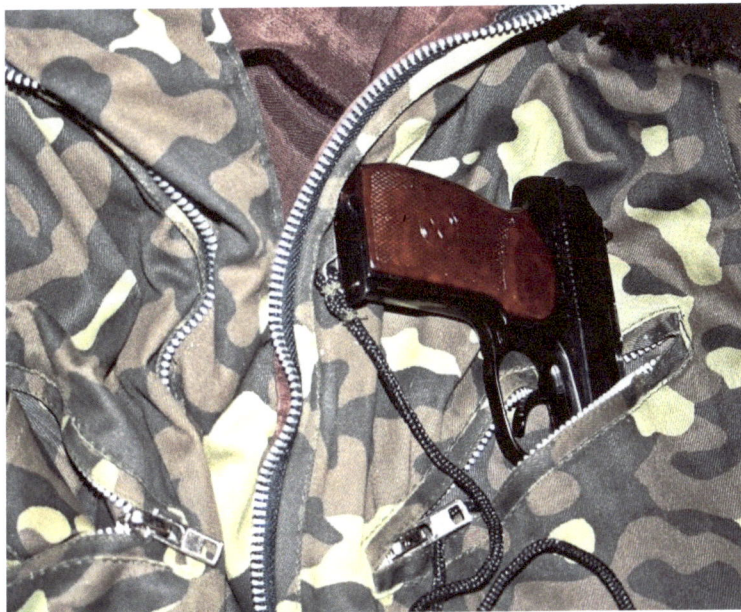

Soviet jackets issued to specialists such as aircrew, armoured crew and officers often featured built in holsters, lanyards, and magazine pouches.

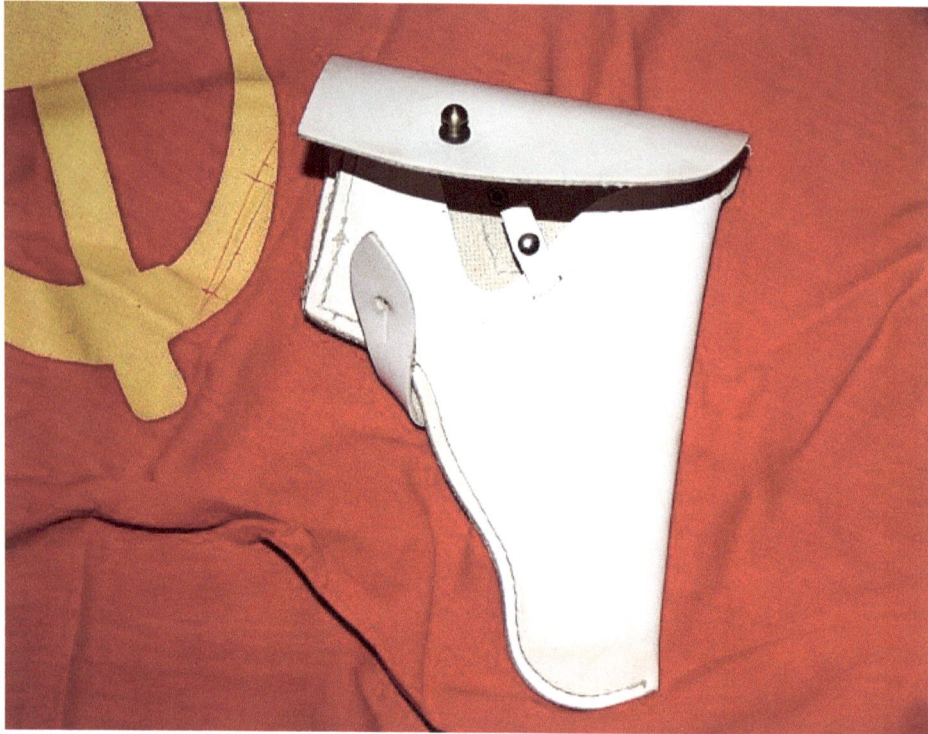

White vinyl parade holster.

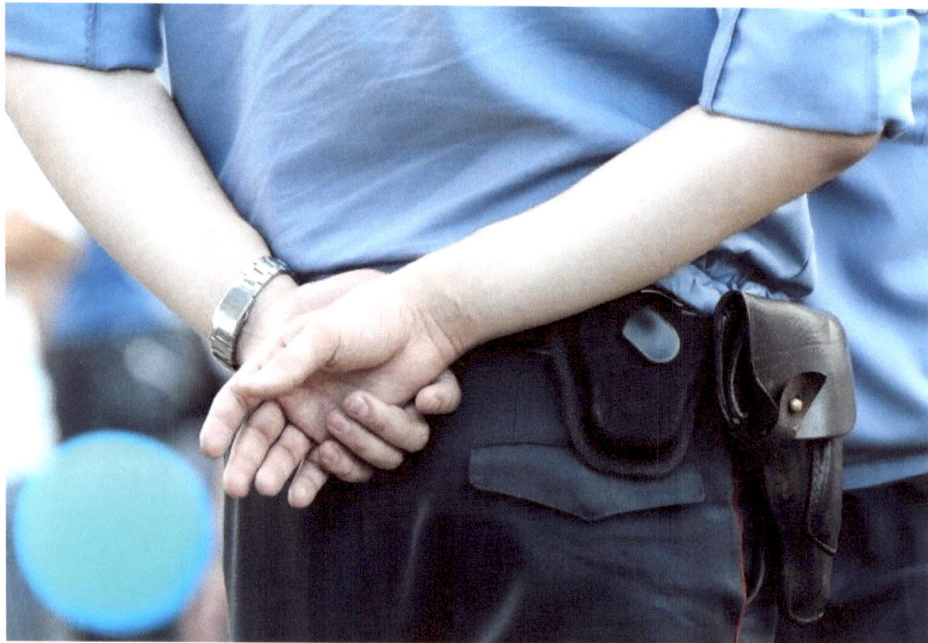

Russian policeman with holstered Makarov pistol.[50]

Post-Soviet Holsters

SPOSN KP-42 Cordura holster.

SPOSN KO-11 Cordura holster.

EFA – 2 self-cocking holster.

Izhevsk leather holster.

Cleaning Rods

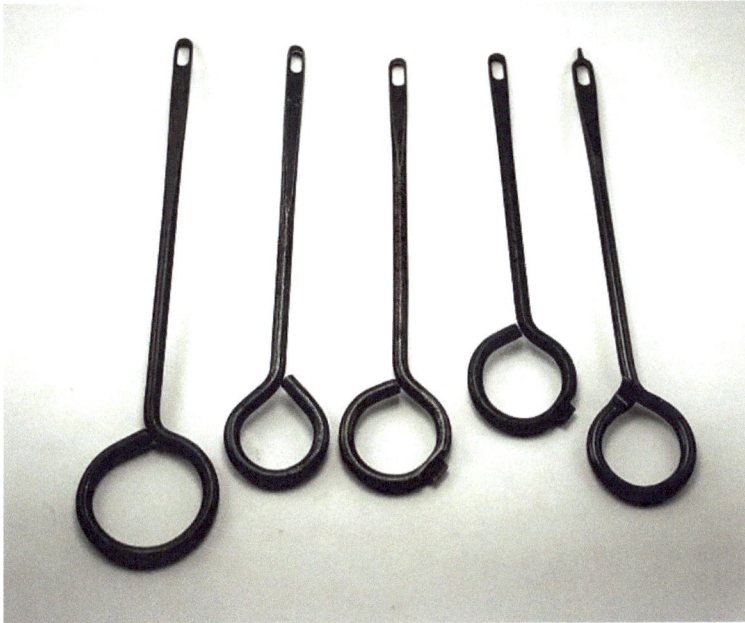

Cleaning Rods (left – right)

TT-33 cleaning rod; early Makarov cleaning rod; early Makarov cleaning
rod with added screwdriver blade; shorter variation of cleaning rod with
screwdriver blade; final (and most common) Makarov cleaning rod with
screwdriver blade in new location, and pointed tip to aid in disassembly.

Cleaning rod screwdriver location on early production cleaning rod (left)
and later production (right).

Lanyards

Lanyards (top – bottom)

Late brown leather lanyard with nickel plated fittings; early lanyard with stitched construction; black leather Navy issue lanyard with brass fittings.

Attachment of pistol, lanyard and waist belt.

9x18mm Makarov Cartridge Specifications

9x18mm Cartridge Specifications

Cartridge weight:		10 g (0.35 oz)
Bullet weight:	Steel Core:	6.1 g (92.9 gr)
	Lead Core:	6.07 g (93.7 gr)
Bullet diameter:		9.2 mm (0.363 in)
Bullet jacket:		Bimetal - steel plated with gilding metal
Propellant weight:		0.25 g (0.01 oz)
Case length:		18.1 mm (0.71 in)
Case material:		Brass, lacquered steel, polymer coated steel or plated steel
Overall length:		25 mm (0.984 in)
Case neck diameter:		9.8 mm (0.386 in)
Case rim diameter:		9.8 mm (0.386 in)
Maximum chamber pressure:		1200 kg/cm^2 (17068 psi)

Bullet Designations

P	(П)	Lead Core
PPT	(ППТ)	Tracer
BZhT	(БЖТ)	Armor Piercing
PBM	(ПБМ)	Armor Piercing, aluminium jacket
RGO28	(РГ028)	Enhanced Penetration
SP-7/SP-8	(СП7/СП8)	Reduced Penetration
PPE /BSZ	(Ппэ)	Hollow Point

Cartridge Designations

57-N-181	(57-Н-181)	Lead Core Bullet, Brass Case
57-N-181S	(57-Н-181С)	Steel Core Bullet, Steel Case
7N25	(7Н25)	Armor Piercing Bullet, Steel Case
57-N-181UCh	(57-Н-181УЧ)	Dummy, Steel or Brass Case with two grooves on case

Cartridges and Head Stamps

57-H-181C Cartridge showing sectioned steel core bullet with steel case and Soviet Army manual.

Soviet 9x18mm cartridge head stamps (left-right)
'270' Lugansk Factory, Ukrainian SSR, brass case
'539' Tula Factory, Russian SFSR, steel case
'38' Yuryusen Factory, Russian SFSR, steel case.

Dummy cartridge, plated steel case. '38' Yuryusen arsenal head stamp.

Soviet military issue 9x18mm in original 16 round box.

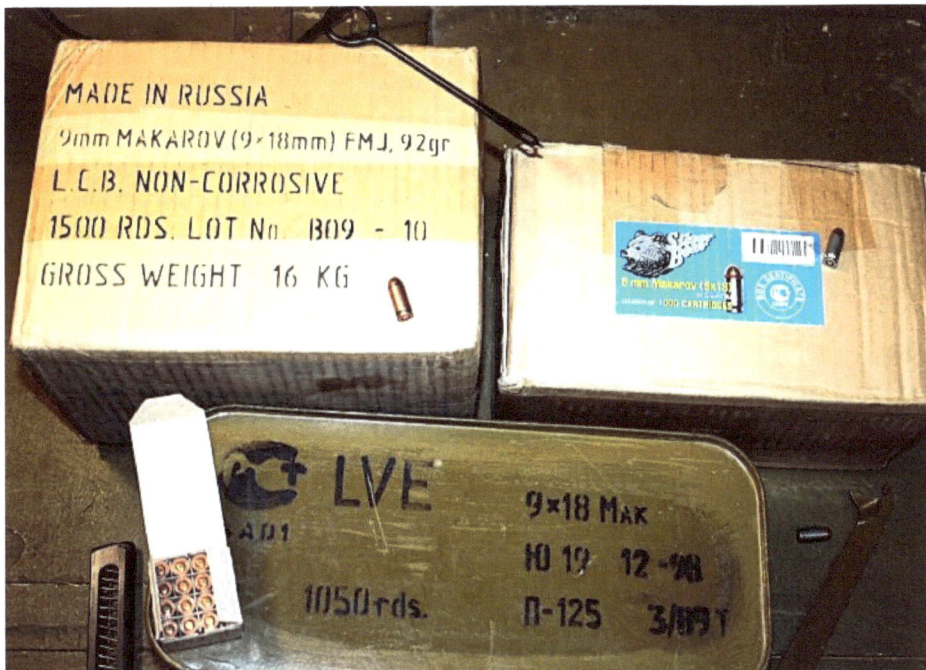

Various commercial bulk packaging of 9x18mm. The metal 'spam can' is identical to its military counterpart.

Training Pistols

Occasionally Makarov pistols are seen marked with the Russian word for 'training' or 'practice' on the slide, УЧЕБНЫЙ or УЧ. These pistols may have one or more 'windows' cut away to allow the interaction of the components to be directly observed. Pistols with MMr or MAKET marked on the slide are products manufactured for commercial sale. These are mass and dimension models of pistols which have various cutaway areas and modified components rendering them totally incapable of firing any ammunition.[51]

Contemporary Soviet 9x18mm Pistols

The armed forces of the Soviet Union fielded two other 9x18mm caliber pistols contemporaneously with the Makarov. The Stechkin Automatic Pistol (APS) is a large, select fire pistol issued to specialists units. The rigid wooden or Bakelite holster also doubled as a detachable shoulder stock. Its magazine contained twenty cartridges in double columns. A later rework, the Automatic Silent Pistol (APB), featured a detachable wire shoulder stock and suppressor. Other than the shared caliber there was no commonality of parts with the Makarov.[52]

Stechkin Automatic Pistol (APS).[53]

Suppressed Pistol (PB).[54]

The Suppressed Pistol (PB) is another specialist pistol. Purposely built with an integral silencer, it was closer in size to the Makarov. The PB also shared the same magazine and a few parts with the Makarov, but it remains an entirely different design.[55]

Manuals

Manual of Instructions for Use and Maintenance for 9mm Makarov Pistol (1955). Succeeding editions showed only slight changes, but the quality of the manual declined.

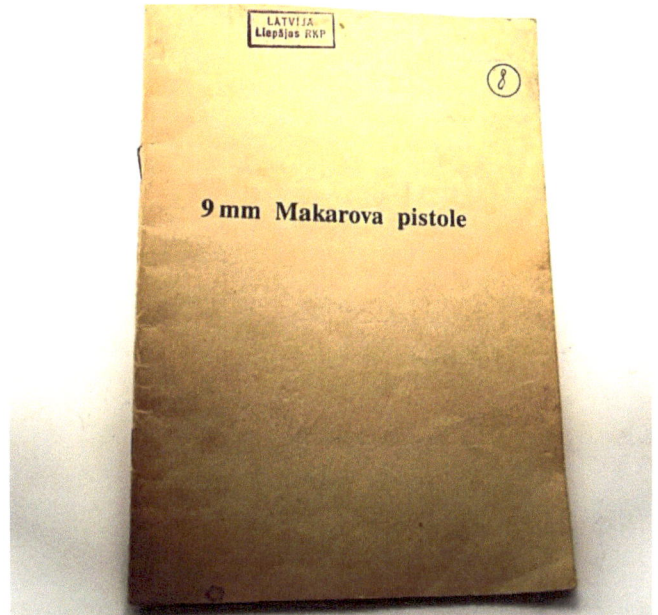

Latvian manual, *9 mm Makarova pistole* (1992).

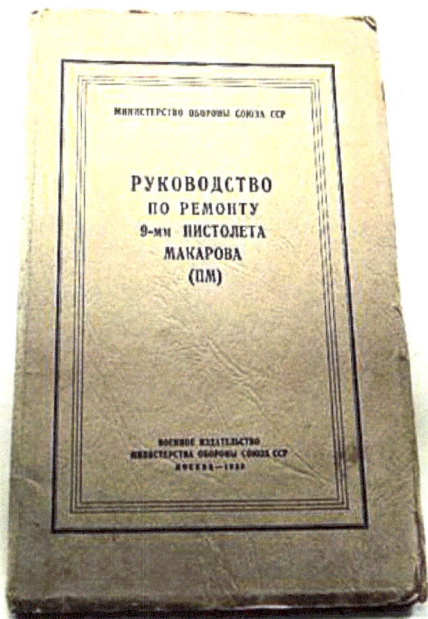

9mm Makarov Pistol (PM) Handbook for Repairs (1956).

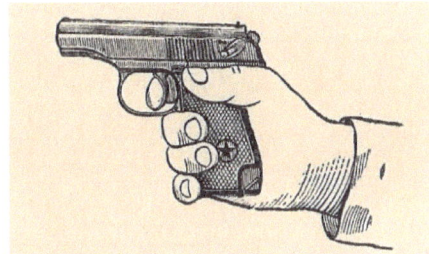

Manual of Instructions (1955) diagram for firing the pistol.

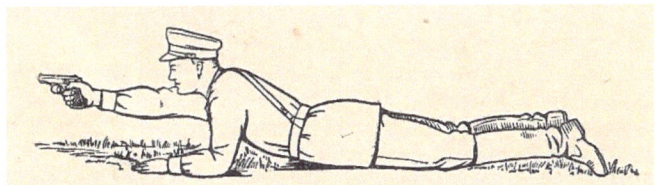

An example of a firing command in the *Manual of Instructions* (1955): "From the prone position, at the deserter, fire." [56]

Ephemera

Pistol rack. Both wooden and plastic versions have been observed.

Ephemera (left to right)

Makarov keychain; N.F. Makarov bronze medallion, ceramic 'Makarov' vodka decanter and shot glasses; exquisite 1/6 scale model of Makarov; Makarov coffee mug; polyurethane Makarov movie prop.

Makarov pistol chocolate.[57]

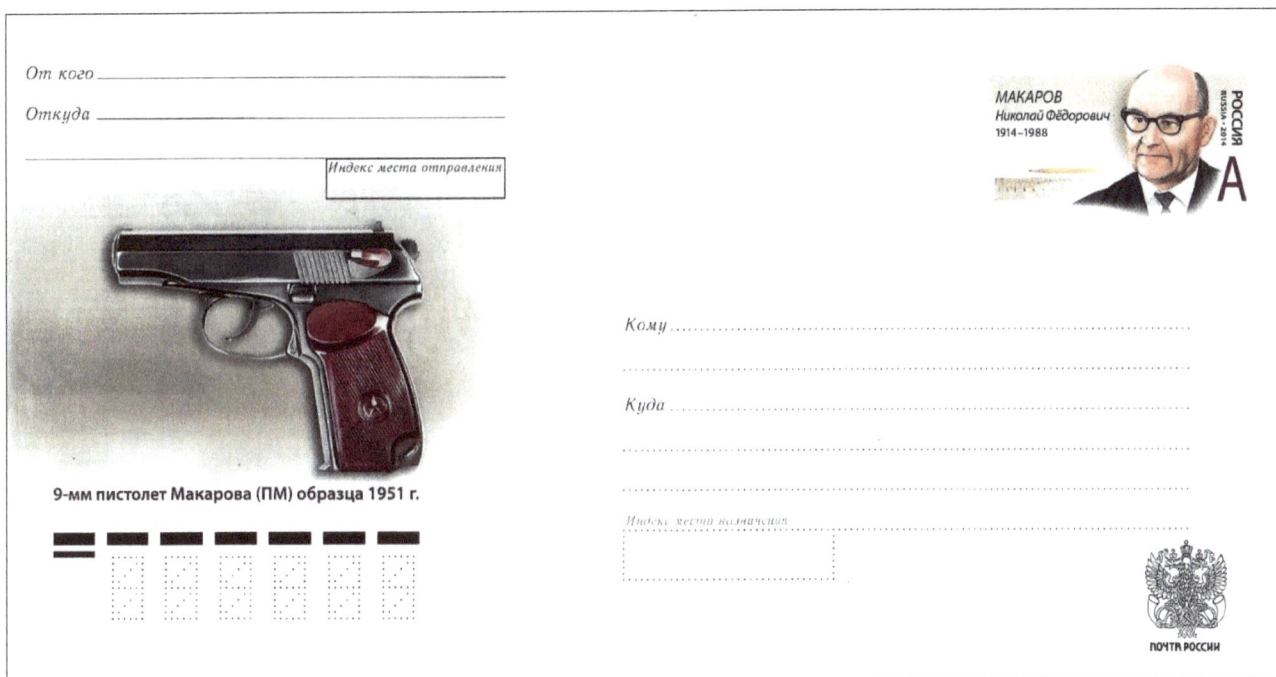

От кого _____

Откуда _____

Индекс места отправления

9-мм пистолет Макарова (ПМ) образца 1951 г.

МАКАРОВ
Николай Фёдорович
1914–1988

РОССИЯ
RUSSIA • 2014

А

Кому ....

Куда ....

Индекс места назначения

ПОЧТА РОССИИ

Nicolay Fyodorovich Makarov and his pistol were honored in 2014 with a limited edition stamped commemorative cover issued by the Russian postal service.

Importation into the United States and Import Markings

The importation of military and police Makarov pistols manufactured in the Soviet Union or its successor state, the Russian Federation, into the United States is prohibited. Indirect importation of Soviet or Russian military Makarovs has occurred after they have been mixed in with shipments of Makarovs from other countries. Technically, all imported Makarov pistols should bear markings indicating the country of origin, but this is not always true.

There are three indirect routes by which Soviet or Russian made Makarov pistols could have entered the United States. Bulgaria purchased and used a number of Soviet Makarov pistols prior to licensed production. Often dubbed by collectors as 'sneaks', such pistols bear all the normal Soviet markings with the additional 'Made in Bulgaria' import stamp.[58]

The German Democratic Republic purchased and used Soviet Makarovs to supplement existing stocks of the locally manufactured 'Pistole M' Makarov pistols. Additionally, when the Soviet military withdrew from the newly re-unified Germany, it left large stocks of weapons behind which, naturally, include Izhevsk made Makarov pistols. These were import marked 'Made in Germany'. The third route was the bringing home of individual 'war trophies', or souvenirs. These do not bear any import markings. A very few of these will have war trophy documents from the Vietnam War detailing the circumstances of the weapon's capture.

Importer Markings Found on Soviet Military Makarov Pistols

CKC	Christopher K. Clark, Portsmouth, Ohio
CAI	Century Arms International, Georgia, Vermont
PW ARMS	PW ARMS, Inc. Redmond, Washington

Importer Markings Found on Commercial Russian Makarov Pistols

KBI	Harrisburg, Pennsylvania
Big Bear Arms	Big Bear Arms, Dallas, Texas
B-West	B-West, Tucson, Arizona
EAA	European American Arms Corporation, Cocoa, Florida (Baikal 654K)

Soviet Makarov with US import marks. Note the 'Germany' marking, indicating that this Soviet pistol had been used in East Germany, and imported from Germany after re-unification.

Two 1981 Makarovs, neither of which bear US import marks. Both show evidence of heavy use. The bottom pistol has mismatched slide and frame serial numbers, something not commonly seen among collector's pistols.

Policemen of the Volkspolizei of the German Democratic Republic aim with the Pistole M.59

9-mm Pistole M

9mm Makarov Pistol

German Democratic Republic

Cameron S. White

Introduction to the Pistole M

The East German Makarov pistol, officially designated the Pistole Makarow or Pistole M (PM), is considered by many to be the quintessential Makarov.[60] This is because the Pistole M was patterned after the early Soviet Makarov and more importantly, because they were built to an unsurpassed standard of quality. The high degree of uniformity and quality was possible because they were produced by skilled workers at one location over a relatively brief period of about eight years. The Pistole M truly represents the unique joining of a proven Soviet pistol design with exacting German production standards. The Pistole M was the primary sidearm used by the various branches of the East German military for over thirty years, which indicates that it served its intended function very well.

Unfortunately, the complete history of the East German Makarov will probably never be known because East Germany was probably the most secretive and repressive communist bloc government ever to exist. Source documents will likely never surface (most were probably destroyed or hidden) and those who built these fine pistols are slowly fading with the passage of time and taking their precious knowledge with them. Despite the lack of information, the story of the East German Makarov is worth piecing together and telling.

Today, the East German Makarov has a small, but dedicated following among collectors and shooters alike. Many of these fans consider the East German Makarov to be the highest quality Makarov ever made, with parts tolerances and bluing equal to or surpassing, commercially produced weapons of the same era. Collectors marvel at the elegant design and how timeless these pistols remain fifty plus years after being made. Shooters like the excellent trigger pull and the inherent accuracy of these pistols due to the fixed barrel design. There is also strong collector interest in the Pistole M based on East Germany itself - the allure of the repressive government and the Stasi, as well as the German arms making tradition.[61] The great irony is that the East German Makarov, in all its perfection and style, was the product of a flawed and short-lived communist state.

The Pistole M is easily field stripped into three major components; the slide, frame and magazine.

Origins of the German Democratic Republic

While a detailed history of the origin and history of the German Democratic Republic or 'East Germany' is beyond the scope of this book, it is helpful to have a basic understanding of the historical and socio-political context in which the East German Makarov existed.[62]

Post-World War 2 Occupation of Germany

East Germany had its origins in the final stages of the Second World War. The Yalta and Potsdam Conferences, which were held by the Allied powers in early and mid-1945 respectively, set the stage for how post-war Germany would be broken into distinct sectors or zones after the war ended. Ultimately, Nazi Germany was defeated and Germany was broken up into four zones; the U.S. Zone, the British Zone, the French Zone, and the Soviet Zone. The states of Brandenburg, Mecklenberg, Saxony (Sachsen), Saxony-Anhelt (Sachsen-Anhelt) and Thuringia (Thuringen) were part of the Soviet Zone and formally became the German Democratic Republic under the communist government of the Socialist Unity Party of Germany in 1949.[63]

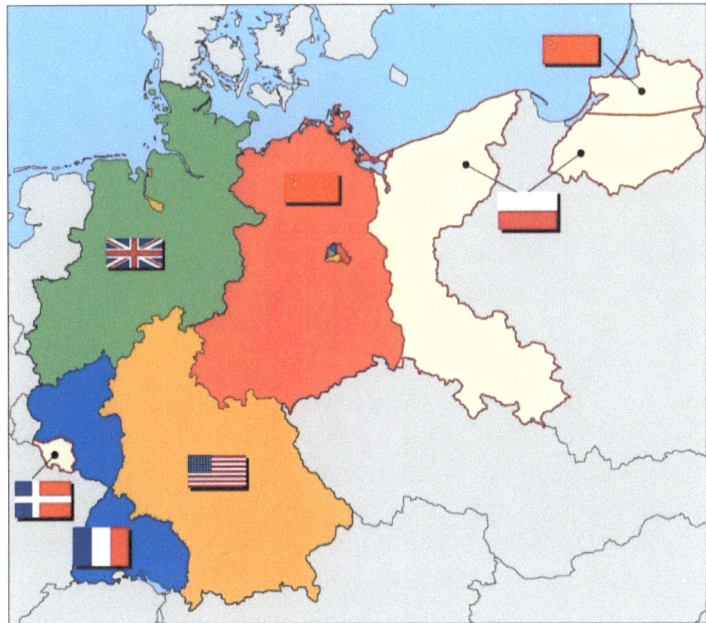

Map showing the occupation of German following WWII. Note that Berlin was divided between the four occupying countries. PMs were produced in Suhl, Thuringia which was part of the Soviet occupation zone, and later, the German Democratic Republic.[64]

Rise and Fall of East Germany

Communist East Germany had a short life span of just over 41 years existing from 7th October 1949 until 30th October 1990. According to census records, about 19 million people lived in East Germany when it was formed, but due to low birth rates and emigration, this number actually dropped to about 16 million by the 1980s. When the country was young, it limped along with the help and direction of the Soviet Union. However over time, the Soviets became less directly involved in East German affairs. During the 1960s and 1970s, various political initiatives aimed at improving the country took place, but the shrinking population was unsettled. The "coffee crisis" of the 1970s, in which regular coffee became unaffordable and was replaced with an unpalatable state-provided blend, was representative of the types of deprivations the population suffered. During the 1970s and 1980s, East Germany's international debt increased, internal protests grew and the country weakened. The fall of the Berlin Wall in November 1989 foreshadowed the total collapse of the communist regime a few months later in March 1990. The country best known for spying on its citizenry fell apart surprisingly quietly and quickly. In October 1990, the German Democratic Republic abolished itself and the states of East Germany reunified with West Germany in the Federal Republic of Germany.

The Makarov is Adopted

It is not clear exactly when the Makarov was officially adopted as the East German service pistol, but it was probably in 1958 or 1959. It was logical for East Germany to use the Makarov, given the strong influence the Soviet Union had on East Germany in the late 1940s and early 1950s. In fact, Makarovs made in the Soviet Union were in use by East Germany in the early 1950s before licensed production began. Many East German officers were trained in the Soviet Union during this time, so they were very familiar with this weapon.

Prior to adopting the Makarov, East Germany utilized a potpourri of WW2- era handguns as their service weapons.[65] For example, the Luger P08, the Tula Tokarev 33 (TT-33), the Walther P38, and the Walther PP were all put into East German service. East Germany even produced a few copies of Luger and the PP on their own machinery. The Pistole M gradually replaced the other handguns as the primary service or "structure" weapon, although the older guns continued to be used on a more limited basis for a long period of time.

It is worth noting that the mid to late 1950s was the right time for East Germany to start manufacturing the Pistole M. The German arms making tradition was still vibrant and many skilled workers still lived in East Germany which provided a ready work force. Manufacturing facilities that expanded during WW2 were available, although a great deal of the machinery in them had been taken back to the Soviet Union for war reparations. The political climate was even encouraging. For example, the Makarov was adopted during the Second Five Year Plan of the Socialist Unity Party of Germany, which ran from 1956 to 1960. The slogan for the Five Year Plan was "modernization, mechanization, and automation". While the focus of the plan was on heavy industry, it is easy to see how the adoption of the PM fit into the zeitgeist of the times.

Pistole M Specifications

Caliber:	9x18mm Makarov (bore diameter 9.2 mm or 0.363 in)
Barrel:	4 grooves, right hand twist, chrome lined
Barrel length :	93 mm*
Weight with empty magazine:	730 g*
Weight with full magazine:	810 g*
Pistol length :	161 mm*
Pistol height :	127 mm*
Magazine Capacity:	8 rounds*
Practical rate of fire:	30 rounds per minute*
Range:	50 meters*
Sighted in distance:	25 meters*
Type of action:	Blowback; single and double action
Firing pin:	Inertia type
Number of parts:	25; 26 (grip with grip screw detent bushing)
Number of slide serrations:	10 left side, 17 right side
Material:	All parts are made of solid steel except the grip (note: the grip has a steel reinforcing strip moulded in the back of the grip)
Durability of pistol:	5,000 rounds [66]

* Specifications from the 1972 East German military manual for the Pistole M.[67]

Pistole M Production at VEB Ernst Thälmann-Werk Suhl

East Germany was the first country outside of the Soviet Union to adopt the Makarov pistol. The Pistole M was made under license from the Soviet Union at VEB Ernst Thälmann-Werk in Suhl, Thuringia, East Germany from about 1957 until 1965.[68] The city of Suhl has a rich tradition of weapons production going back to the 1500s. Many well-known firearms firms were based in Suhl including Haenel, Krieghoff, Merkel, Sauer, and Simson. It is not surprising then, that the Pistole M was made in a city with a strong arms making tradition. VEB Ernst Thälmann-Werk Suhl was actually the successor of parts of the Simson firm and other firearms firms which had long been in existence. Apparently, VEB Ernst Thälmann-Werke Suhl was a conglomeration of several facilities. Unfortunately, little is known of how VEB Ernst Thälmann-Werk Suhl was formed, the production process or the workers who made them. The Weapons Museum in Suhl indicates that production of the Pistole M began in 1957 and ended in 1965 at the request of the Soviet Union.[69]

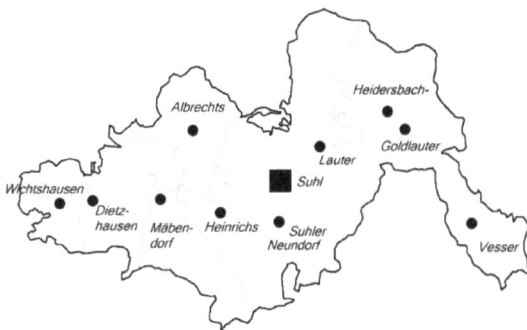

The city of Suhl in the state of Thuringia, Germany.[70]

The Simson firm, portions of which eventually became VEB Ernst Thälmann-Werk Suhl, had a long and interesting history that is worth detailing briefly, since it is so closely associated with the Makarov saga.[71] The Simson family, which founded the firm, had long lived near Suhl, Thuringia. The roots of the Simson company began with the initiative and business acumen of brothers Loeb and Moses Simson. They were originally cattle dealers, but branched out into other businesses including iron forges and gun parts manufacture in the 1850s after laws affecting Jewish business ownership were liberalized. The Simson firm prospered and continued in the firearms business throughout the latter half of the 18th century. Around the turn of the 19th century, the firm diversified and began producing bicycles, automobiles, swords and bayonets as their main products. Simson then returned to small arms during WWI when they began producing the Gewehr 98 and machine gun parts and other war materiel. During the early 1920s the Simson firm was awarded lucrative government contracts for P08 production and other small arms work which sustained the company through the early 1930s. Things changed dramatically in the middle 1930s when the firm was taken over by the Nazis who ruled the country. After much harassment, the owners decided to escape to Switzerland and eventually the US, but the factory was turned over to the Nazis as part of an agreement. An era had ended. The company was then renamed Berlin-Suhler Waffen (BSW) and later to Gustloff Werke in honor of Nazi leader Wilhelm Gustloff, who was assassinated in 1936. The Gustloff plant produced a tremendous amount of armaments during WW2 and was a huge sprawling complex.

At the end of WW2, United States troops occupied parts of Thuringia for several months, however this area ultimately fell into the Soviet Occupation Zone. After Soviet troops moved in, they destroyed some of the Simson buildings and sent much of the machinery from the Simson factory to the Soviet Union. What was left of the factory was used to produce strollers, motorcycles and some firearms. Firearms production was then transferred to VEB Ernst Thälmann-Werk Suhl, which was one part of the original factory. Surprisingly little information has surfaced about the VEB Ernst Thälmann-Werk Suhl for the period that Pistole M was made there. After the reunification of Germany, the plant was again renamed and it became Suhler Jagd-und Sportwaffen GmbH (Hunting and Sporting Weapons Suhl). Simson Suhl Makarovs were made at this renamed plant in the mid-1990s. The firm declared bankruptcy and permanently closed in 2003.

Early Production 1957-1959

The Weapons Museum in Suhl reports that prototypes of the East German Makarov were made in 1957. 1958 was probably a set up or pre-production period, since no guns dated 1958 have been observed by collectors, although pictures have been reported. The start of true production occurred sometime in 1959, but the early production methods were found to be problematic. One source states that parts were die cast which resulted in breakages.[72] Reportedly the plant managers were imprisoned for the problems with production. Manufacturing methods changed due to the early problems and parts went back to being milled which resolved the issues.

Late Production 1960-1965

The early production problems were worked out and pistols were produced continuously until October 1965 by which time several hundred thousand had been produced.[73] There are no mechanical differences between early and late production pistols.

The Pistole M appears to have been the primary service pistol in use up until the time that East Germany dissolved in 1990. A few pistols were used by the state of Sachsen-Anhalt after reunification and were marked 'LSA' (Landespolizei Sachsen-Anhalt) on the frame. The 'LSA' marking was struck out with lines when the pistols were sold as surplus. Pistols with the 'LSA' marking are considered to be scarce and are only seen occasionally.

'LSA' marking below the slide stop on a 1960 Pistole M, serial number H 0394.

LSA marking on a 1961 Pistole M, serial number AW 5487. Note variation in font for 'LSA' marking.

Production Numbers

Without official production records, it is hard to say exactly how many East German Makarovs were made, however the number is almost certainly in the hundreds of thousands. This estimate is based on the idea that each prefix code represents a run of 10,000 pistols. It has been noted that 170,000 pistols were shown on an inventory roster in East Germany in 1989, so at least that many were produced.[74]

Serial Numbers and Markings

The Pistole M has no markings to identify its model or country of origin, however it is easily identified and distinguished by its markings and grips. The serial number markings on the left rear of the frame provide a great deal of information. The serial number and the year code were marked on the frame with a series of electro-pencilled dots as opposed to being machine engraved. This results in the serial number being rather light and hard to read.

Each pistol has a one or two letter prefix before the serial number. Pistols made in 1959 and 1960 have a one letter prefix and most pistols made 1961 through 1965 have a two letter prefix. It is not known why certain letters or letter combinations were chosen for the prefixes, especially for 1958 through 1960. However, it does appear that the letters followed a rough alphabetical progression from A, B, D, E to F from 1961 to 1965. The date and associated prefix codes are as follows:

Year	Associated Prefix Codes [75]
1958	S
1959	J, K, L, N, U
1960	B, C, E, F, G, H, M, T
1961	AR, AS, AQ, AT, AU, AV, AW, AX, AY, AZ, D
1962	AO, AP, BR, BS, BT, BU, BV, BW, BX, BY, BZ
1963	BN, BO, BP, DA, DB, DE*, DF, DK, DL, DP
1964	ES, ET, EV, EW, EX, EZ
1965	ER, FB, FD, FF, FH, FJ

*A single pistol with a DE prefix and a 64 year date has been observed. This pistol appears to be an anomaly.

All pistols have a four digit serial number a few spaces after the alphabetical prefix. For each unique prefix, the serial numbers appear to run from 0000 to 9999. Serial numbers ranging from 0044 to 98xx have been observed by collectors. While the full range of serial numbers has not been observed by collectors, enough have been noted in the ranges to make the reasonable assumption that there are 10,000 pistols per block.

A two digit date code follows the serial number. The date codes observed are '59', '60', '61', '62', '63', '64' and '65', with each number representing the last two digits of the year the pistol was produced. Therefore, a pistol with the date code of '64' would have been produced in 1964.

There is often a stamping after the date code or below the serial number. There is much speculation as to what these markings represent, but these are believed to be inspection markings. There are a wide variety of markings ranging from what appears to be letters including 'a', 's', 'y' and 'z', to what appears to be a backwards 's' and a lower case 'r'. There are often punch marks or dots around the markings, but there is little certainty as to their meaning. Some have suggested that each mark represents part of the inspection process. Nearly all pistols also have a small marking of a rectangle with 'K100' inside of it. All pistols also have a small square marking with a dot inside it on the frame and slide that is believed to be an inspection mark. There are occasionally other manufacturing or assembly markings on other parts such as the slide.

Distinguishing Characteristics

Grips

The Pistole M typically had dark black checkered grips unique to East German Makarovs. The grips have no distinguishing markings or external symbols and appear to be made from a hard plastic, possibly Bakelite. The grips have a piece of reinforcing metal moulded on the inside surface of the back strap to provide strength. There are two basic patterns of Pistole M grips; an early pattern and a late pattern.

Early Pattern
There are two variants of the early grip; one that has fine checkering and one that has medium checkering. Both variants of early grips used the Soviet style grip screw with a fluted head, and the grip screw hole was drilled deeper than the later pattern to accommodate the thicker screw head. The early grips had the Soviet style grip screw detent bushing in the screw hole. The early grips do not have any manufacturer markings on the inside surface. Very few of the early grips are seen, which probably indicates they were used for a relatively short period of time compared to the later grip.

Late Pattern
The late pattern grips used a grip screw with a wider diameter head which was thinner than the early pattern grip screw. The screw hole of the late pattern grip was not bushed. It is not known why the change was made to a different type of grip screw, although it may have been a manufacturing simplification by the elimination of the bushing.

Late grips often have a marking moulded on the inside bottom rear of the grip. The symbol is a very small triangle with rounded corners inside a circle with the letters 'DAMW' inside it with the numbers 'N40' and '71' outside the triangle. The number '3568/5' or '3568/6' is usually found directly above the triangle and are larger in size than the symbol. Other grips only have the number '3' moulded in the bottom rear section. The parting mark on the outside rear of the grip is a prominent groove on some grips, whereas it is much less noticeable on other grips.

Marking on inside of late grip.

Early Soviet type grip screw (left), later GDR grip screw (right).

It is also worth noting that while the typical color of early and late grips is jet black, a very few of the late pattern grips are a chocolate brown color or have red, green or brown flecks in them. These grips are seldom seen and are highly prized by collectors. The reason for the color variation has not been explained. The brown grips appear to contain some type of fiber, as opposed to being solid plastic so perhaps they were experimental.

Early Pattern Grips

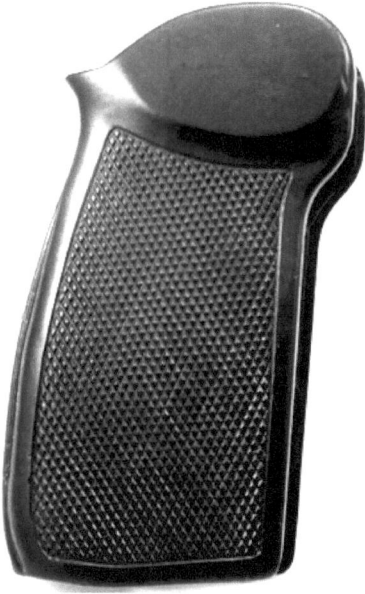

Black grip, early pattern, with fine checkering.

Black grip, early pattern showing grip screw hole with detent bushing and hole for lanyard.

Black grip, early pattern with medium checkering.

Late Pattern Grips

Black grip, late pattern.

Rear of late pattern grip.

Rare brown grip.

Lanyards

Many grips had an extra hole drilled in the back to accommodate a lanyard. This modification was recommended in 1969 due to troops losing pistols. Lanyards were 100 cm long and black or green in color. The lanyard was attached in one of two ways, either with a metal clip that was fitted on the inside the grip, or simply knotted behind a lanyard hole drilled into the grip. The drilled lanyard hole is often off center or slightly out of round which gives the impression that the work was done hastily.

Black lanyard with metal fitting attachment.

Close up of metal fitting attachment.

Magazines

The magazines used in the Pistole M appear to be similar to Soviet ones of the same era in that they have a hump on the bottom of the left side. However, the top of the hump is rounded on East German magazines, whereas the Soviet magazines are flat. Some Pistole M magazines have the last four digits of the serial number of the pistol electro-pencilled on the bottom of the left side of the magazine, parallel to the floor plate. It is quite unusual to find a pistol with a matching magazine. The majority of magazines are seen without serial numbering, and many of these unnumbered magazines are probably replacements. Magazines have been seen with a number of different markings. For example, some have been observed with an additional single number, usually '4' or '5', stamped on the left side of the magazine body. Other markings that have been observed include the number '2' stamped on the spine, and a small lower case 'g' or 'h' stamped on the left side of the magazine body. The significance of these numbers and letters is not known. One magazine observed had fine checkering on the follower projection, but this is not common.

Magazines

Right side of typical magazine.

Rear view.

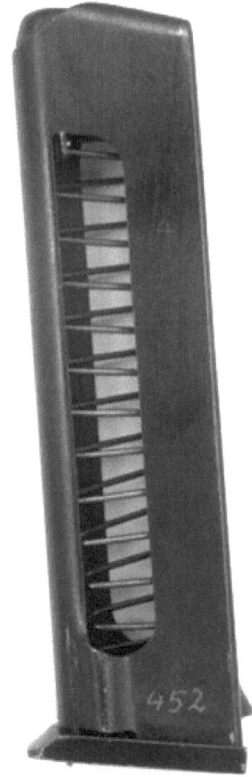

Left side of magazine. Note the electro-pencilled final digits of the serial number.

Magazine Markings

Small 'g' stamp on magazine body.

'5' stamp on magazine body.

'2' stamp on magazine body.

Policeman of the Volkspolizei demonstrates the complete
Pistole M assembly of frame, slide, magazine and lanyard.[76]

Collector's Review of the Pistole M (1957-1965)

There appear to be few variations in Pistole M pistols other than external markings and the pistol remained remarkably unchanged during its short production run. This section list details, observed differences, and collector comments by year of production.

1957 & 1958 Pistole M

Observations
There are no known collector's examples of the earliest production pistols. Pistols made in 1957 and 1958 were pre-production samples or prototypes and most were probably replaced or destroyed over time due to parts breakage.

Serial Number Prefix
S [77]

Estimated Production
The production was likely very small - probably only a few dozen prototype, pre-production, and tool room samples were made.

Variations
No known variations.

Collector Comments
1957 and 1958 dated pistols are the most highly sought after East German Makarov, given that there are no known collector's examples. Collectors live in hope of finding an extant specimen.

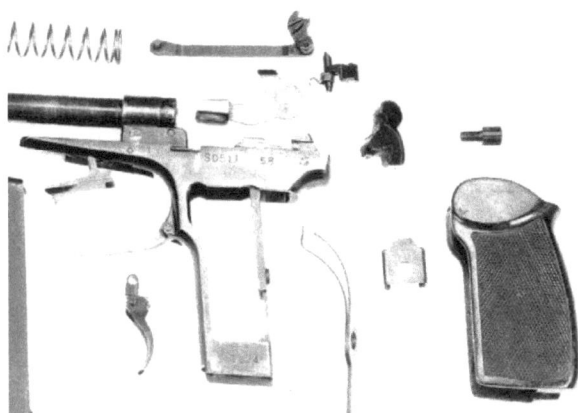

Only one photograph is known of either a 1957 or 1958 Pistole M. A 1958 Pistole M, serial number S0511 with the early pattern grip and grip screw, is shown in *Small Arms of the World*, 12th Revised Edition by Edward Clinton Ezell. Note the '58' date code and the inspection mark.[78]

1959 Pistole M

Observations

1959 represents the first year of regular Pistole M production, although it is not known exactly what part of 1959 that production started. 1959 dated pistols are rarely encountered and collectors estimate that there are only about 10 or 15 examples in the United States. The reason for the extremely low number is probably due to manufacturing problems that likely resulted in early pistols being either recalled or destroyed. Some parts were initially made from castings which resulted in breakage. Eventually, all parts were machined.

Serial Number Prefixes

J, K, L, N, U

Estimated Production

There were probably far fewer than 50,000 produced and many were undoubtedly destroyed due to parts breakages. There appears to be a very low survival rate for these pistols.

Inspection Marks

The square inspection mark is below the serial number on the slide (instead of behind the safety as was the convention for other years). The pistol pictured has an 'a' inspection mark on the right of the serial number.

Variations

A variation in the manner in which the serial number is applied has been noted. Some pistols, probably earlier production, have Soviet style engraved serial numbers on the frame and slide, while others are marked using a series of electro-pencilled dots. Due to the limited number of collector's examples, it is difficult to document other differences.

Collector Notes

Based on its rarity, a 1959 dated Pistole M would be the centerpiece of a Makarov collection. 1959 pistols are very occasionally seen for sale. One came up for sale in 2012. One that was offered at an online auction site in 2015 sold for over $1,300.

Left side of 1959 Pistole M serial number N 2391. Note that the 'circle in square' inspection mark is near the serial number on the slide.

Left side of 1959 Pistole M serial number K 5856. The serial number was applied by stamping or engraving, similar to early Soviet Makarovs.[79]

Left side of 1959 Pistole M serial number N 2391.

Right side of 1959 Pistole M serial number N 2391.

1960 Pistole M

General Observations
1960 is the last year for single letter serial number prefixes. 1960 dated pistols are often found for sale. They often have more wear than pistols made in other years, probably because they were in service longer.

Serial Number Prefixes
B, C, F, G, H, M, T

Estimated Production
70,000

Inspection Marks
One pistol with a double inspection mark has been observed. Pistols are marked with a 'y', 'z' or a backwards 's'.

Variations
None noted.

Collector Notes
1960 pistols are often seen with the early pattern Soviet style grips with a fluted grip screw.

Left side of 1960 Pistole M, serial number H 0968. Note the 'z' inspection marking.

Left side of 1960 Pistole M, serial number H 0394. Note different font on the slide and the scarce 'LSA' marking below slide stop indicating the pistol was used in post-reunification Germany.

Left side of 1960 Pistole M, serial number H 0968.

Right side of 1960 Pistole M, serial number H 0968.

1961 Pistole M

General Observations
1961 dated pistols are often seen for sale.

Serial Number Prefixes
AR, AS, AQ, AT, AU, AV, AW, AX, AY, AZ, D

Estimated Production
110,000

Inspection Marks
Pistols are marked with a 'y' or a symbol that resembles a lower case 'a' or a backwards 's'.

Variations
None observed

Left side of 1961 Pistole M, serial number D 4154. Note 'y' inspection mark.

Left side of 1961 Pistole M, serial number AY 7117. Note the inspection marking that resembles a lower case 'a' or perhaps a backwards 's'. Also note the extremely rare matching magazine.

Left side of 1961 Pistole M, serial number D 4154. Note the D
serial number prefix which is uncommon for a 1961 dated pistol.

Right side of 1961 Pistole M, serial number D 4154.

1962 Pistole M

General Observations
1962 dated pistols are the most commonly observed in the United States. It appears that more 1962 dated pistols were imported relative to other years.

Serial Number Prefixes
AO, AP, BR, BS, BT, BU, BV, BW, BX, BY, BZ

Estimated Production
110,000

Inspection Marks
Pistols are marked with a 'y' or what resembles a backwards 's'.

Variations
The font for the serial number on the slides of AP prefix pistols is different in terms of style and size compared to other pistols. Two pistols in the AP series have been observed by the author and both showed this difference.

Left side of 1962 Pistole M, serial number BS 7184. Note the non-standard 'y' marking that appears to have been stamped more than once.

Left side of 1962 Pistole M, serial number AP 1232. Note the non-standard serial number marking on the slide. The style of the numbering on the slide does not quite match the serial number on the frame.

Left side of 1962 Pistole M, serial number BS 7184.

Right side of 1962 Pistole M, serial number BS 7184.

1963 Pistole M

General Observations
Based on observed specimens, it appears that relatively fewer of these pistols were imported into the U.S. or perhaps for some reason, fewer survived. These come up for sale much less often than other years.

Serial Number Prefixes
BN, BO, BP, DA, DB, DE, DF, DG, DK, DL, DP

Estimated Production
110,000

Inspection Marks
Pistols are marked with a 'y'.

Variations
None noted.

Collector Notes
Many do not realize the relative scarcity of 1963 dated pistols, so they usually sell for the same amount as pistols dated to other years. In terms of relative collector's value, these are a bargain.

1963 Pistole M, serial number DF 7248.

Left side of 1963 Pistole M, serial number DF 7248.

Right side of 1963 Pistole M, serial number DF 7248.

1964 Pistole M

General Observations
The bluing on 1964 pistols appears to be deeper and richer than that of earlier pistols.

Serial Number Prefixes
EL, ES, ET, EV, EW, EX, EZ

Estimated Production
70,000

Inspection Marks
Pistols are marked with a 's', 'y', and what resembles a lower case 'r'.

Variations
None noted.

Collector Notes
1964 is one of the lowest production years for the Pistole M.

Left side of 1964 Pistole M, serial number EX 3971. Note 's' inspection mark.

Left side close up of 1964 P Pistole M, serial number EZ 0044. Note smaller 'r' inspection mark. Note also that this is the lowest serial number observed by collectors in any series.

Left side of 1964 PM, serial number EZ 3283.

Right side of 1964 PM, serial number EZ 3283.

1965 Pistole M

General Observations
1965 was the last year of Pistole M production, with production ceasing in October of that year. The finish is usually a beautiful deep blue, much like 1964 pistols. The last serial number prefix appears to be FJ.

Serial Number Prefixes
ER, FB, FD, FF, FH, FJ

Estimated Production
60,000

Inspection Marks
Pistols are marked with what resembles a lower case 'r'.

Variations
None noted.

Collector Notes
Relatively fewer pistols were made in 1965, but they show up for sale fairly often. They represent a good value when priced comparably to an earlier production pistol. The condition of 1965 dated pistols seem to be generally better than those of other years. Only one pistol has been observed by the author with a FJ prefix.

Left side of 1965 Pistole M, serial number FD 5622. Note the rare brown grips.

Left side 1965 Pistole M, serial number FD 5622.

Right side of 1965 Pistole M, serial number FD 5622. Note the attached lanyard.

Collector's Review of the Simson Suhl Makarov (1994)

General Observations

The Simson Suhl Makarov is not a military pistol, however it is based on the Pistole M and was probably made using some military parts. The Simon Suhl Makarov was first produced by Suhler Jagd-und Sportwaffen GmBH (Suhl Hunting and Sport Arms Ltd) in 1994. Based on serial numbers and reference sources, it appears that between 700 and 1000 were made. An unknown number of these pistols were imported into the United States by Classic Distributers Incorporated (CDI). They were sold for a brief period in 1997 by Southern Ohio Gun Distributors. Simson Suhl pistols that were imported came in a small plastic box with a spare magazine and an owner's manual.

Simson Suhl Makarovs were made in post-reunification Germany for the commercial market and the bluing and fit of these pistols is excellent. The reason for making a small run of these pistols is not known. Perhaps old parts needed to be used up or this was a way for Germany to generate hard currency. There is some debate as to whether left over military parts were used to make these pistols, or whether a combination of old and newly manufactured parts were used. Some have suggested that these pistols contain some Bulgarian made parts, but this has not been proven. The pistols have the same external appearance as the Pistole M, but there was one significant design change. The hammer was altered so the slide could be racked when the safety was on. This is advantageous because it enables a chambered round to be cleared while the safety is on.

Marks

Markings include the CDI import mark, the Simson Suhl mark, and German commercial proof marks.

Estimated Production

It is estimated that between 700 and 1,000 Simson Suhl pistols were produced.

Variations

There are two different types of slide markings. Early pistols with serial numbers in the 300s or lower were marked 'Simson Suhl Thur', while later pistols were marked 'Simson Suhl'.

Collector Notes

Despite being produced thirty years after the last Pistole M was made, Simson Suhl Makarovs command a great deal of collector interest and they are prized by their owners. Based on the low number produced, they are rarely seen for sale and command high prices, usually $1,000 or more for a specimen in excellent condition. Pistols below serial number 500 are seen much less often than those in the 500 to 800 range. No pistol with a serial number above the low 800s has been seen by the author.

Interestingly, the serial number was stamped on the right side of the frame, which is opposite to that of the military models.

The following serial numbers have been observed by the author: 000576, 000628, 000645, 000716, and 000792. Philip Duane Neel tabulated and reported the following Simson Suhl serial numbers: 00050 ('Thur' marked), 000318 ('Thur' marked), 000465, 000518, 000537, 000556, 000580, 000608, 000609, 000640, 000641, 000663, 000670, 000713, 000809 and 000810.

Left side of Simson Suhl Makarov, serial number 000670.

Right side of Simson Suhl Makarov serial number 000670.

Legend on left side of slide of Simson Suhl Makarov, serial number 000670.

Close up of markings on right side of Simson Suhl slide. Note the word 'Suhl' below the anvil.

Right side of Simson Suhl showing serial number and markings on the frame and slide. The alphanumeric code indicates a 1997 production date. Beneath the code is the Suhl proof house mark of a shoe sole and pick (left) and the German proof mark (right).

Close up of CDI import marking.

Gray and blue boxes for Simson Suhl pistols. The box held the pistol, a spare magazine and the operator's manual. The serial number of the pistol was handwritten on the small white label in the lower corner of the outside of the box.

Holsters

Holsters are a commonly seen accessory. Many were imported into the United States at the same time the pistols. Most Pistole M holsters were made of leather and had a compartment on the side for a spare magazine. None had loops to hold a cleaning rod. Early versions of the holster were black and had the closure strap sewn to the top flap, whereas later versions had the closure flap secured to the lower portion of the holster. Most leather holsters were reddish brown, dark brown, black or white. Additionally there was a cloth holster in the Nationale Volksarmee Strichtarn camouflage pattern. It is difficult to make definitive determinations about what group used a holster based on its color. However in general, it is believed that the brown or reddish brown versions were used by the army, the black ones by the riot police and the white holsters were used by the traffic police and for parade purposes. The cloth holsters in Strichtarn camouflage, commonly called the 'raindrop pattern', were heavily produced in the late 1980s, but apparently were never officially issued. A shoulder holster was also issued. This holster was made of light leather and it appears to be rather flimsy compared to the other leather holsters. In addition to the sanctioned holsters, many were crafted at the local level so a tremendous number of variations exist. Interestingly, some Strichtarn pattern field jackets had a cloth compartment sewn on the inside to hold a Pistole M.

Many holsters are stamped on the inner flap with 'TGL' followed by a number. The TGL and number is the material standardization number. Many holsters also have ink stamps on the inside of the flap to indicate the organization they were issued to. For example, the marking 'MdI' or 'MdJ' stands for Ministerium des Innern (Ministry of the Interior), responsible for the Deutsche Volkspolizei, the national police force. Some are marked 'MfS' which stands for Ministerium für Staatssicherheit (Ministry for State Security) responsible for the State Security Service or 'Stasi'. Some early holsters are also date stamped in ink on the inside of the flap. Commonly observed year stamps are 1965 and 1966, although earlier dates are seen from time to time. Holsters with the date stamp also often have a Roman numeral 'I', 'II', 'III' or 'IV' which indicates the quarter of the year it was produced.

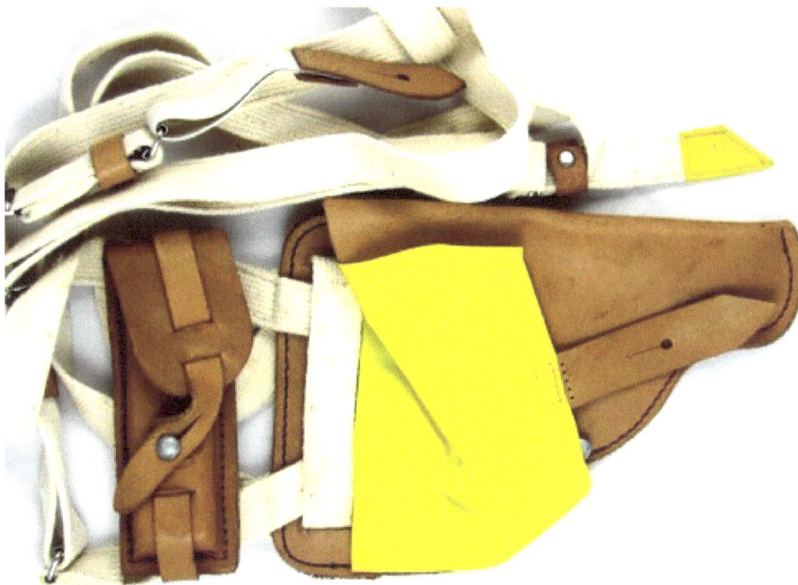

Leather shoulder holster with yellow flap and tag.

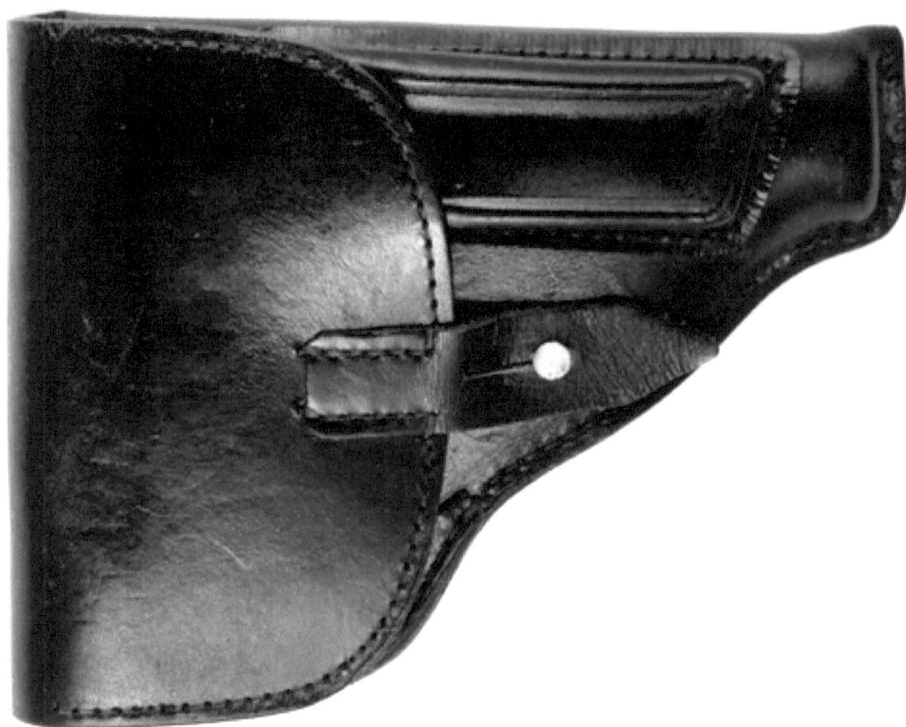

Early pattern black holster.

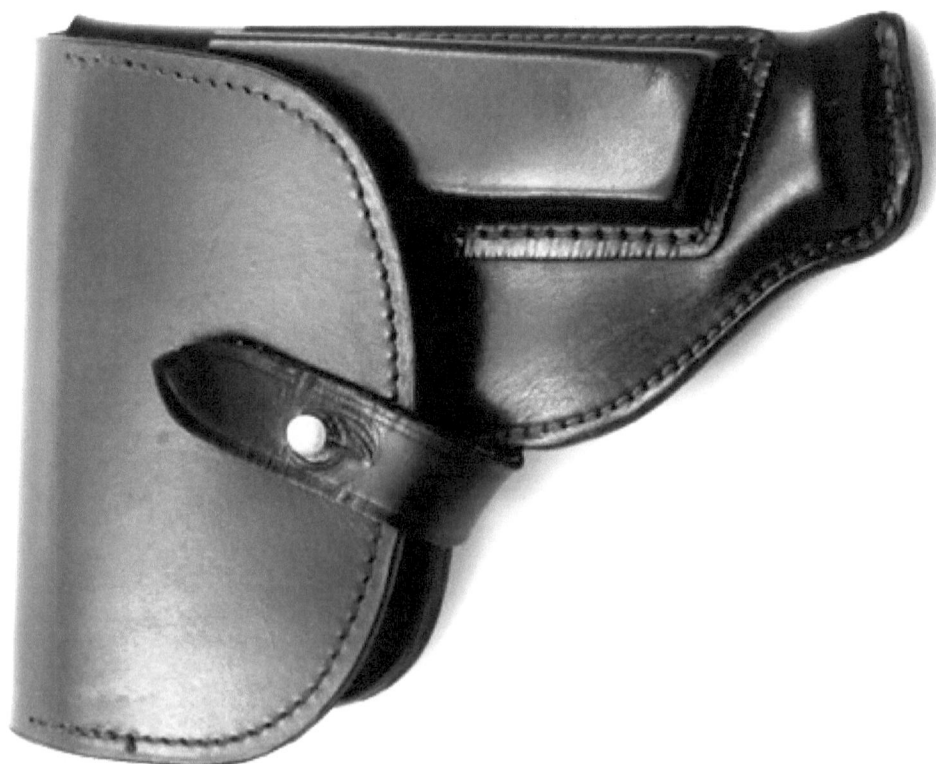

Late pattern black holster.

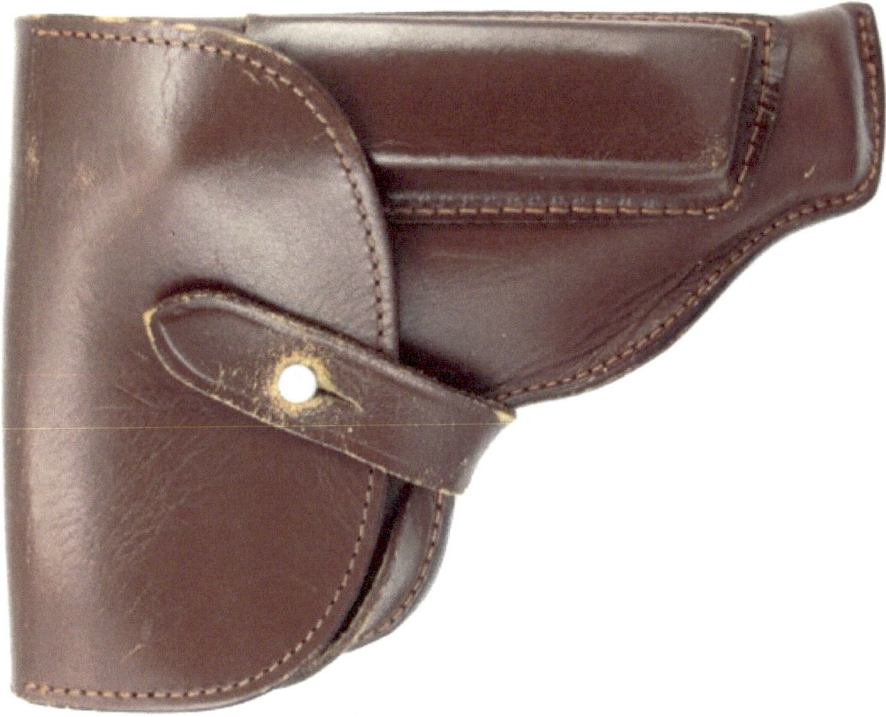

Reddish brown leather holster.

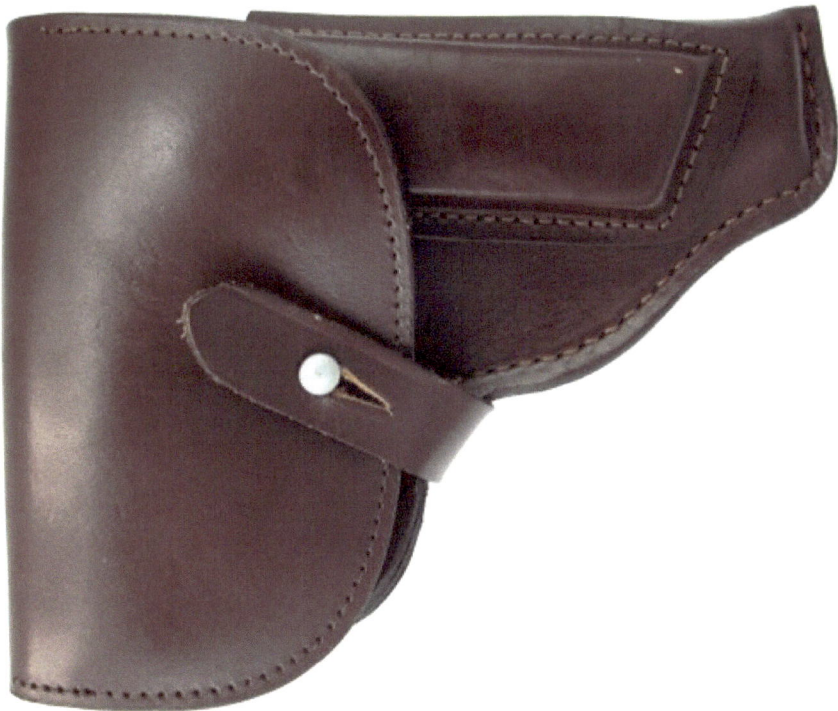

Darker brown leather holster.

Strichtarn cloth holster.

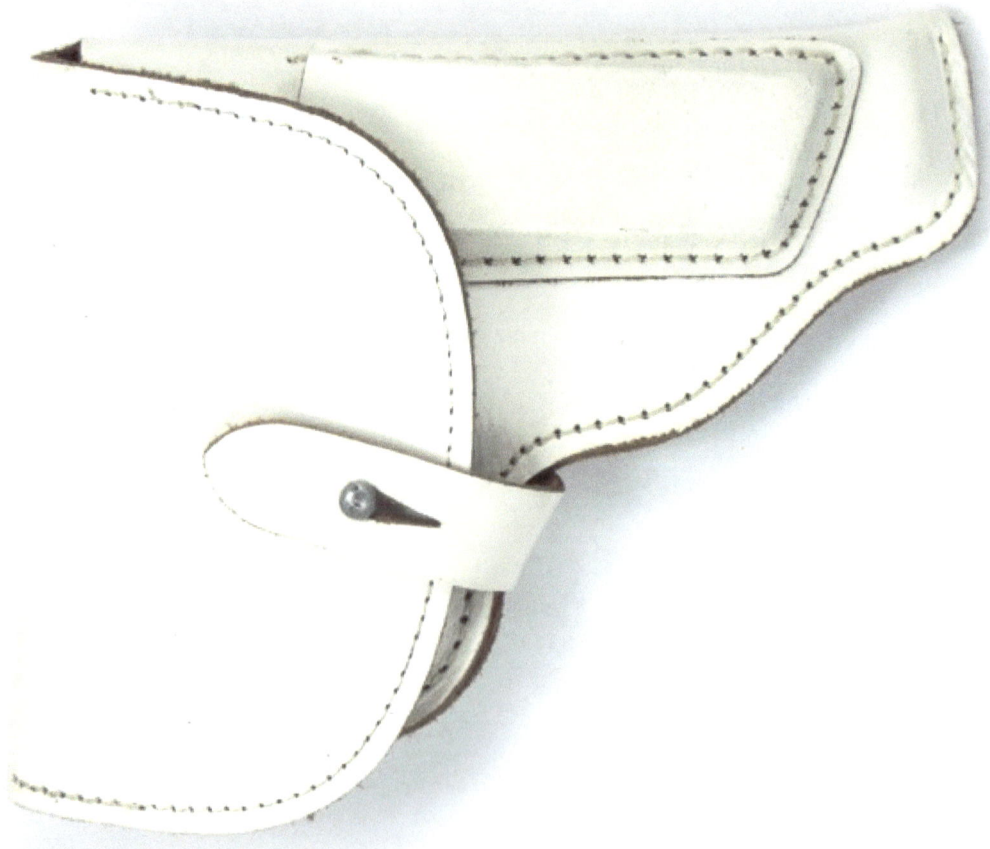

White traffic police and parade leather holster.

'NVA' (Nationale Volksarmee) and
70 date marking inside of holster.

'MDI' (Ministerium des Innern)
marking inside holster. This marking is
the one most commonly seen.

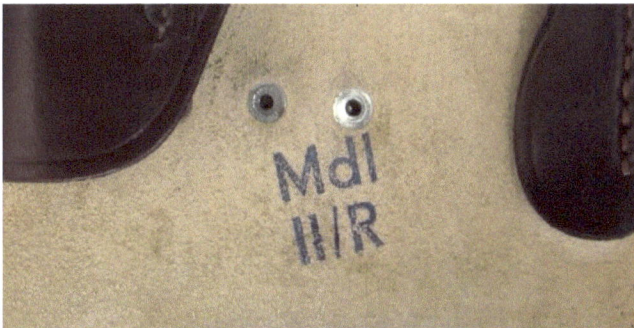

'MdI' marking inside holster.

'MdI' marking inside white holster.

'63' date stamp on inner flap of holster.

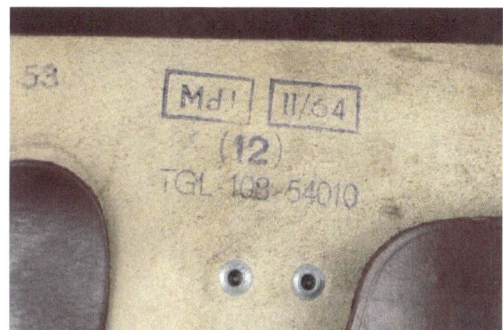

'64' date stamp on inner flap of holster.

'TGL' and '65' date marking inside holster.

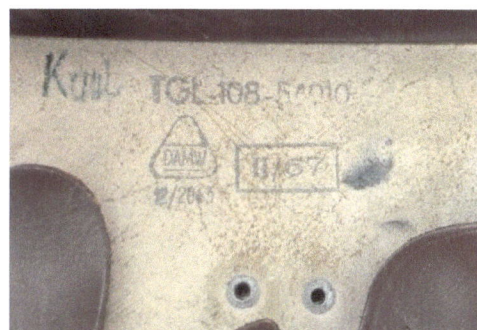

'TGL' and '67' date stamp on inner flap of holster.

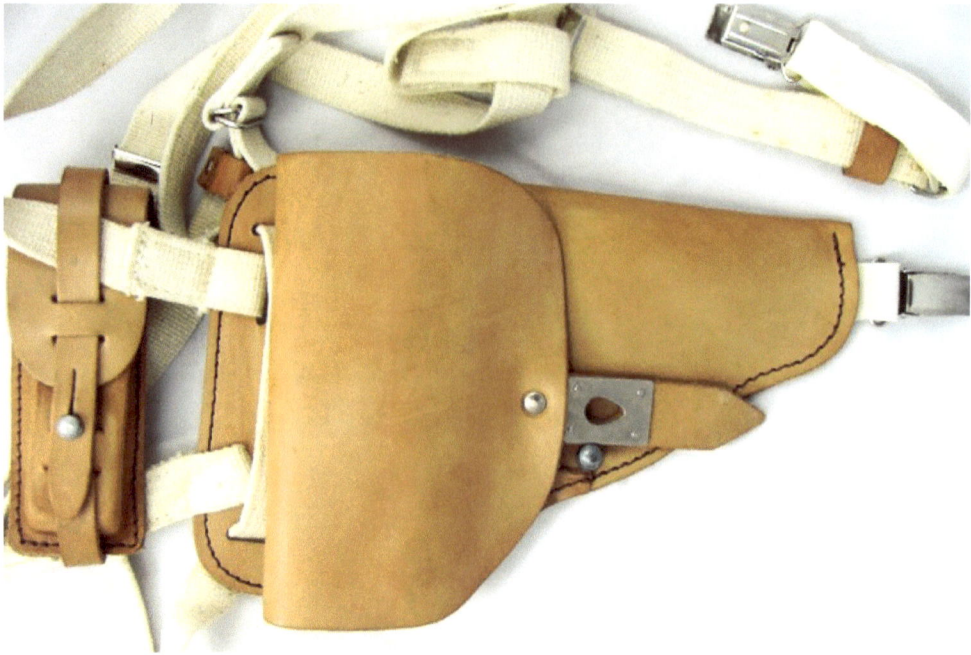

Leather shoulder holster.

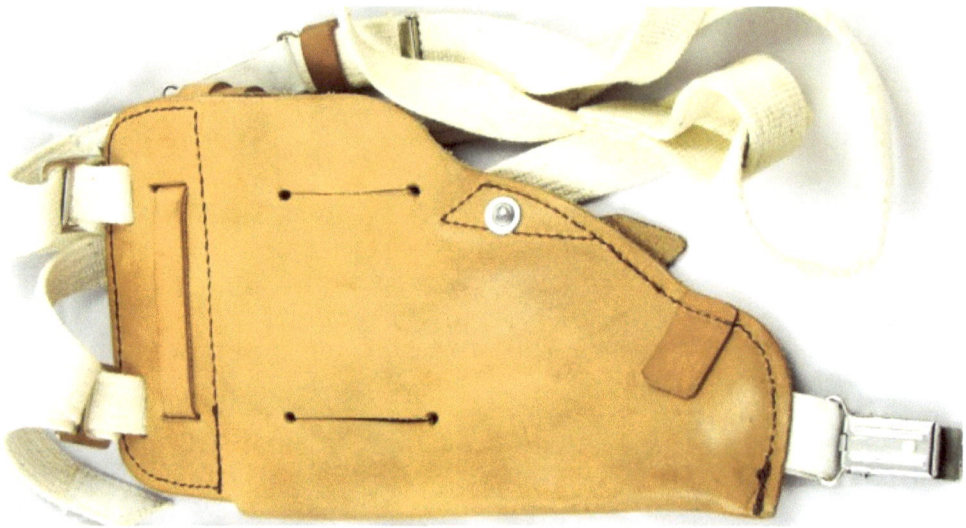

Rear of leather shoulder holster showing cuts for a strap or belt.

MdI' marking on inside of top holster.

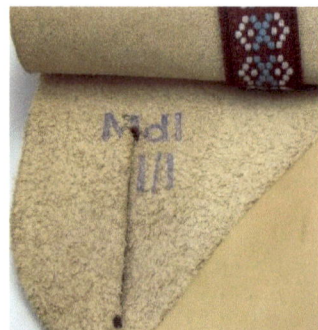

'MdI' marking on inside of bottom holster

Cleaning Kit

The Pistole M was originally issued with a Reinigungsgerät 57 (RG57) cleaning kit, primarily intended for 7.62x39mm rifles, but an universal item for both the AKM rifle and the Pistole M. Later, when East German forces introduced the AK-74 rifle, the Pistole M was issued with an AK-74 cleaning kit, normally contained in a canvas roll, but often retaining the RG57 kit's metal container.[80]

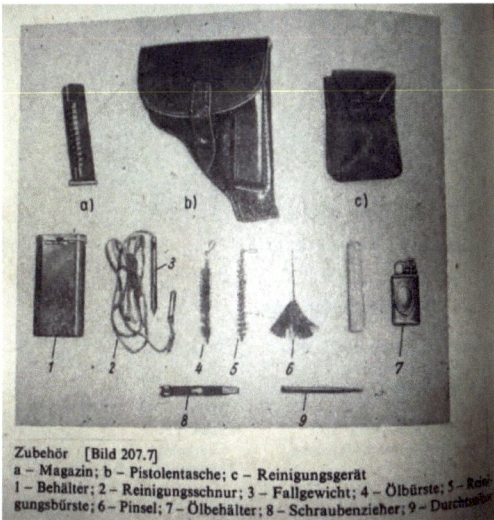

Zubehör [Bild 207.7]
a – Magazin; b – Pistolentasche; c – Reinigungsgerät
1 – Behälter; 2 – Reinigungsschnur; 3 – Fallgewicht; 4 – Ölbürste; 5 – Reinigungsbürste; 6 – Pinsel; 7 – Ölbehälter; 8 – Schraubenzieher; 9 – Durchtreiber

RG57 cleaning kit in the border guards manual, *Handbuch für den Grenzdienst* (1978).[81]

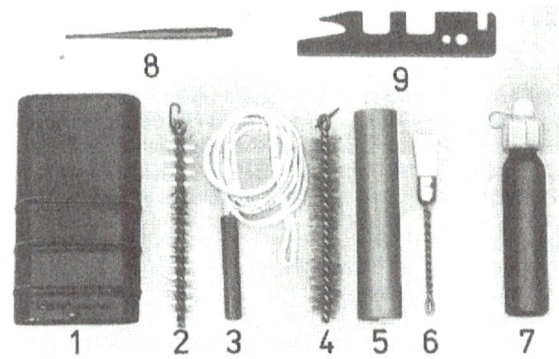

Bild 10 Reinigungsgerät 57 (RG 57)

1 – Behälter; 2 – Reinigungsbürste; 3 – Reinigungsschnur mit Fallgewicht; 4 – Ölbürste; 5 – Schutzhülle für Ölbürste; 6 – Ölpinsel; 7 – Ölkännchen; 8 – Dorn; 9 – Schraubenzieher

AK-74 style cleaning kit in the police manual, *Schießen mit den Pistolen Makarow und Modell 74* (1986).[82]

RG57 cleaning kit.[83]

East German 9x18mm Makarov Cartridges

East Germany produced 9x18mm ammunition for many years. The head stamp on the cartridges is usually '04' followed by the last two digits of the year it was made such as '64', '76' or '79'. The ammunition has a steel case and a steel core bullet with a copper jacket weighing 6.1 g. The muzzle velocity is 315 m/s. Cartridges came in 16 round cardboard boxes with 80 boxes per zinc tin. 9x18mm dummy cartridges were also produced, although they have no identification markings. The cartridge consists of a steel base with a plastic primer insert and a one piece black plastic case projectile.

1965 dated ammo box.

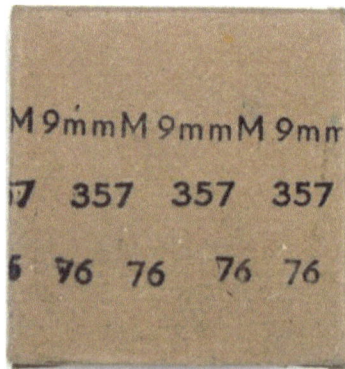
1976 dated ammo box.

1979 dated ammo box.

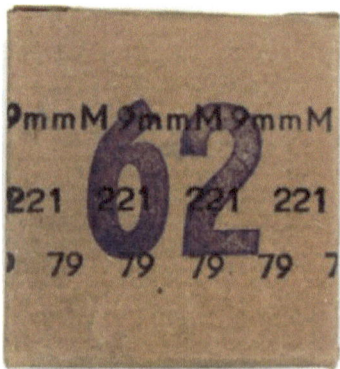
Stamp variation on box.

Sixteen cartridges per box.

Dummy cartridge.

Dummy cartridge base.

1964 cartridge.

1964 head stamp.

1979 cartridge.

1979 head stamp.

Training Devices and Armorer's Accessories

Training Devices

An electric aiming device was developed to assist soldiers learning to shoot. The device was called the Pistolenzielgerat 75 (pistol aiming device) or PZG75. The device was fitted directly on the pistol frame, using front and rear sights on the top of the device for aiming the pistol.

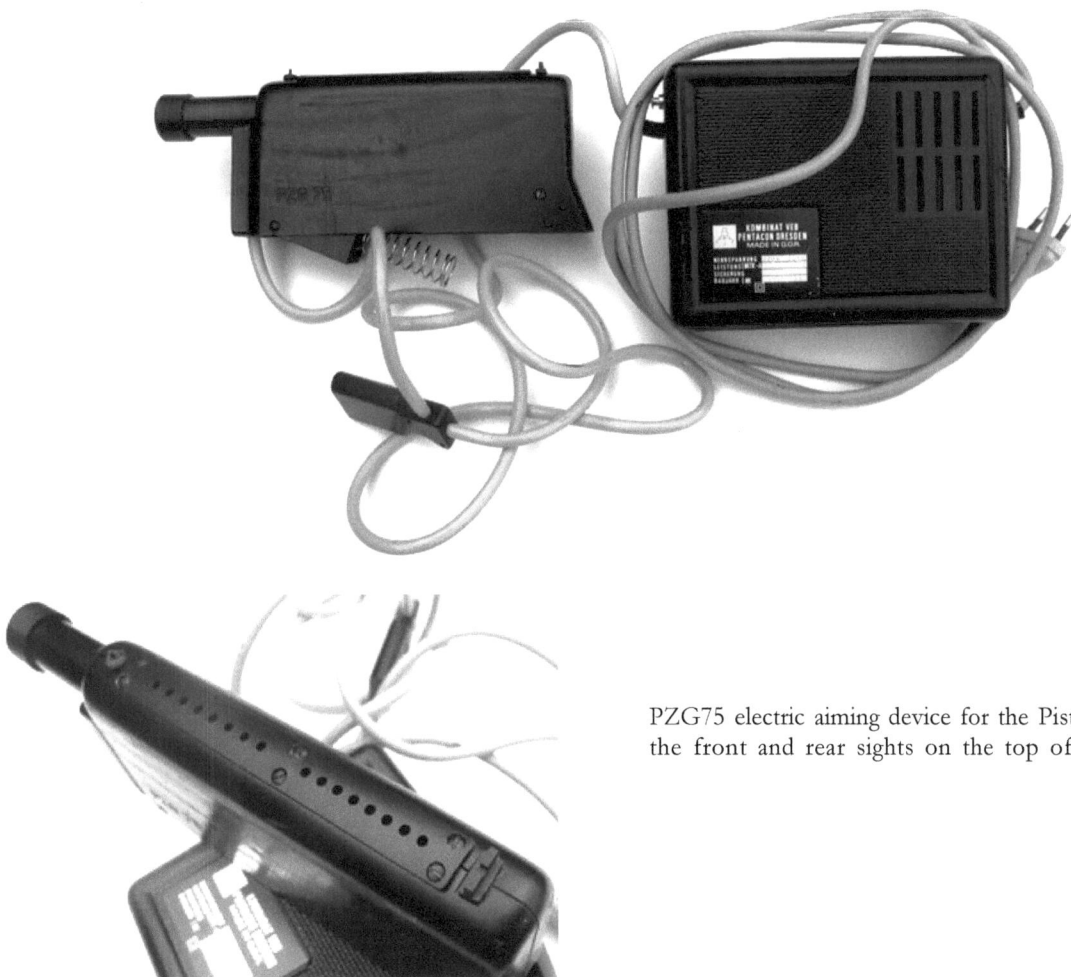

PZG75 electric aiming device for the Pistole M. Note the front and rear sights on the top of the device.

A .22 caliber conversion unit was also used for training. The kit consisted of a slide and magazine.[84]

Some pistols had parts of the barrel, frame and slide sectioned away to expose the inner workings of the pistol and were used for teaching troops about the Pistole M. These are referred to as the *schnittmodell* (cut model) or *lehrwaffe* (instructive weapon) and are often marked 'Lehr' on the slide.

A few engraved presentation pistols were also produced. For example, Erich Honeker, who led East Germany from 1971 to 1989, had a set of engraved pistols that included a Pistole M.[85] A photo of a Simson Suhl Makarov that was plated and engraved is found in Marschall's book.[86] Others undoubtedly exist, but they are exceedingly rare.

Armorer's Spare Parts Box

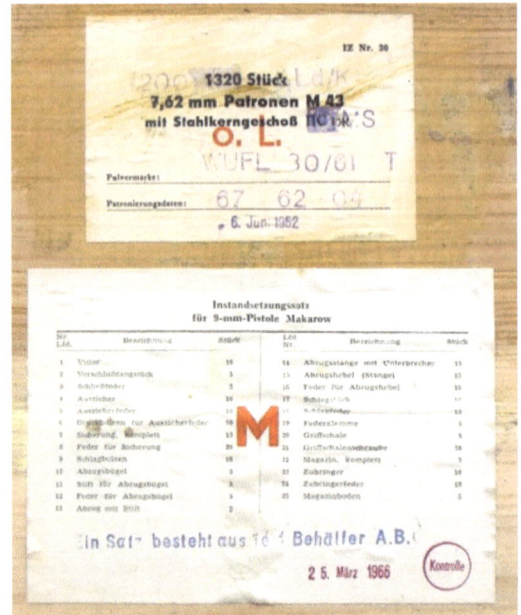

The East German armorer's spare parts box contained a wide range of spare parts ranging from grips to firing pins, but not any major parts such as frames or slides. The parts were often packed in bright yellow containers. Interestingly, the wooden box used to hold the parts was a re-used 7.62x39mm ammunition crate. These parts kits are still available on the US surplus market.

Armorer's spare parts box

Labels on the inside lid of spare parts box

Inside the armorer's spare parts box

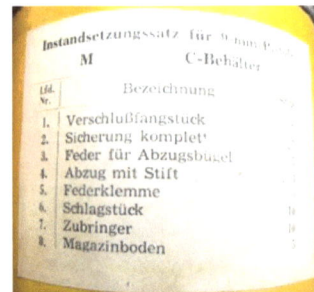

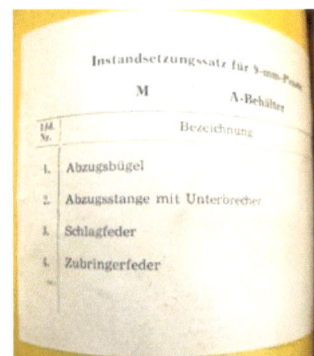

Spare parts canister labels

Armorer's Gauges

The armorer's gauges kit contained several high grade tools that were used to make adjustments or take precise measurements on GDR firearms. The kit contained four gauges used on the Pistole M. The tools are stored in a wooden box that had cut-outs and padding for the tools to protect against damage. A listing of the tools was found on a metal plate on the inside of the lid. The box also contained a record sheet.

Top lid of the armorer's gauge box.

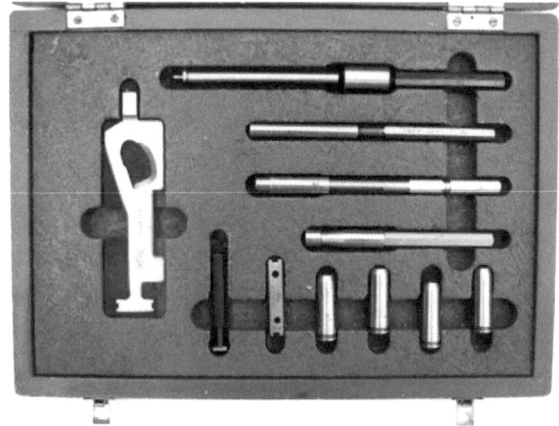

Inside of the armorer's gauge box.

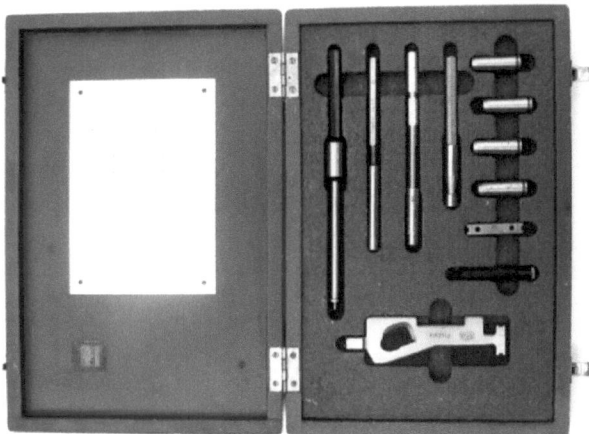

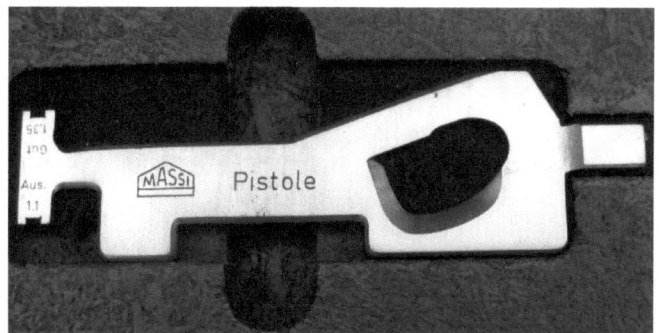

Inside of the armorer's gauge box.

Close up of the gauge for the Pistole M

Metal label on inner lid listing tools and gauges

Record sheet for armorer's gauge set

Manuals and Technical Documents

Two operator's manuals for the Pistole M were published by the Nationale Volksarmee (National People's Army); one dated 1958 and one dated 1975. The 1958 edition is almost never seen by collectors, but 1975 edition is occasionally seen for sale. Several technical manuals for the Pistole M were published by the NVA. One is number A 050/1/442 and is titled *9-mm-Pistole M Befundaufnahme und Qualitatsfeststellung* (1978). The other one also has number A 050/1/442 and is titled *9-mm-Pistole M Pruftechnologie* (1976). A small book published by the Ministerium des Innern (Ministry of the Interior) titled *Schießen mit den Pistolen Makarow und Modell 74* (1986) has some interesting photos and information about shooting the Pistole M.

Log books or maintenance records that matched a particular pistol were included with some of the pistols imported to the U.S. by Coles. These have been seen most commonly with 62 dated guns, but one for a 64 dated pistol has also been observed. The log contains information about the condition of the pistol and the dates it was inspected. The dates recorded in the log books observed begin in the 1980s.

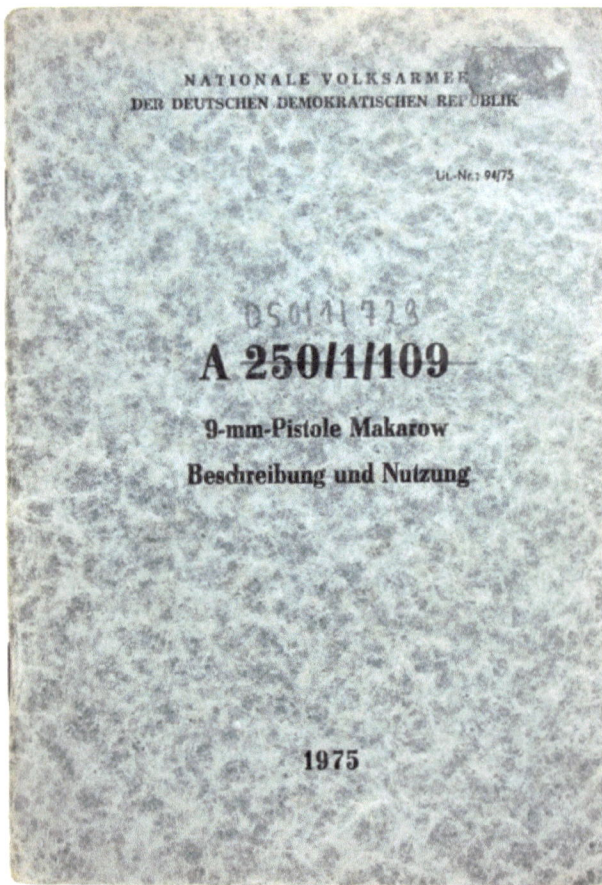

National Volksarmee Pistole M operator's manual, *9-mm Pistole Makarow: Beschreibung und Nutzung* (1975).[87]

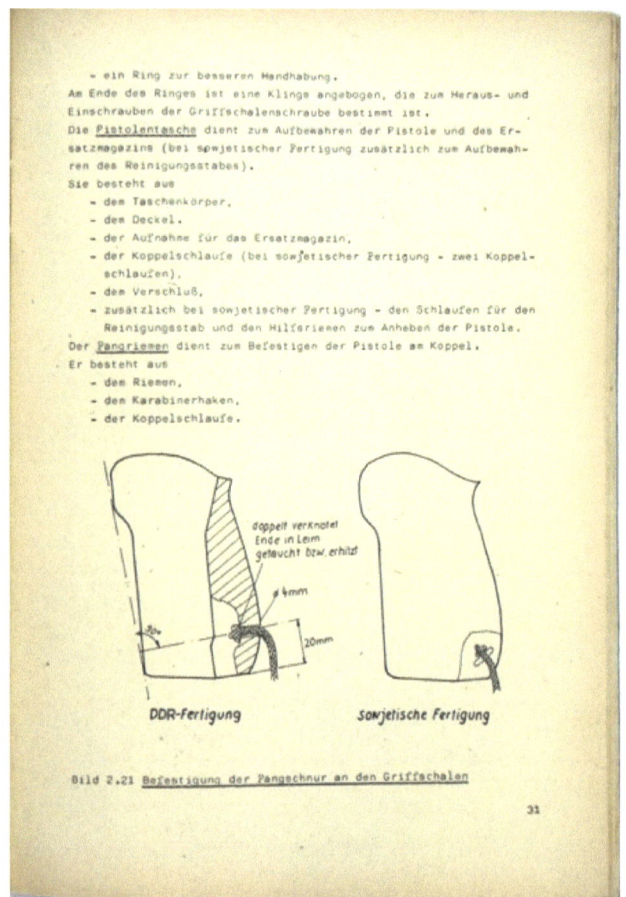

NVA manual showing lanyard attachment on GDR (left) and USSR (right) Makarov grips.[88]

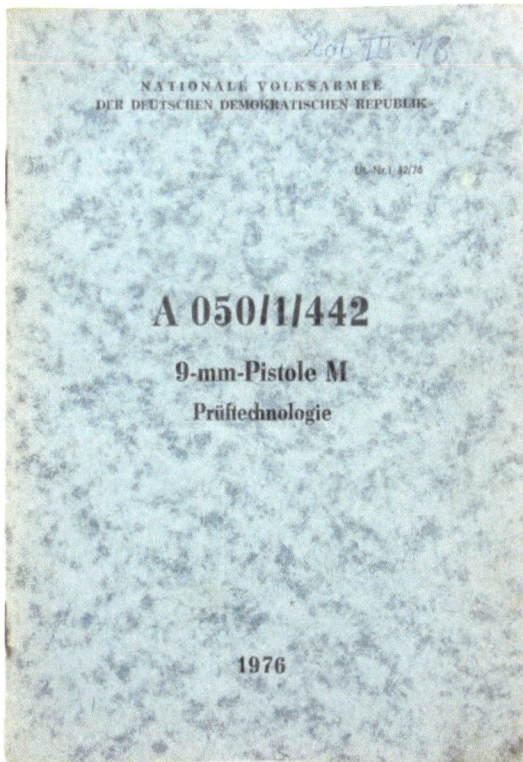

NVA Pistole M technical manual, *9-mm-Pistole M: Prüftechnologie* (1976).[89]

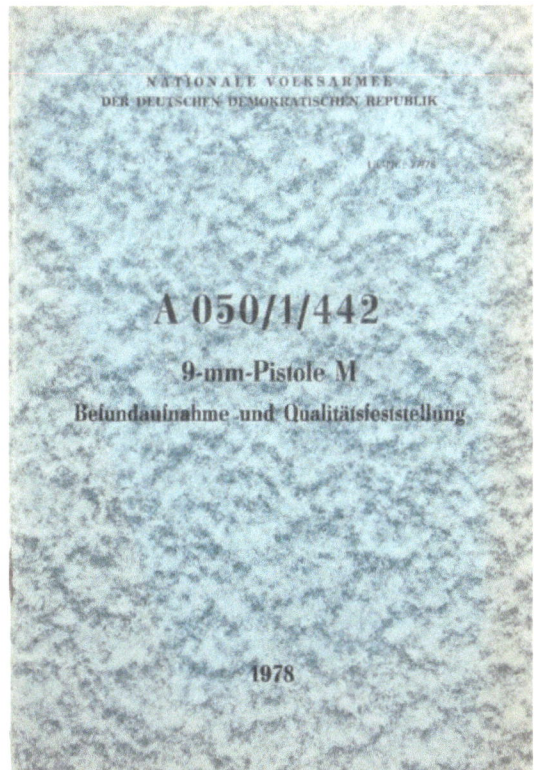

NVA Pistole M technical manual, *9-mm-Pistole M: Befundaufnahme und Qualitätsfeststellung* (1978).[90]

Ministerium des Innern training manual, *Schießen mit den Pistolen Makarow und Modell 74* (1986).[91]

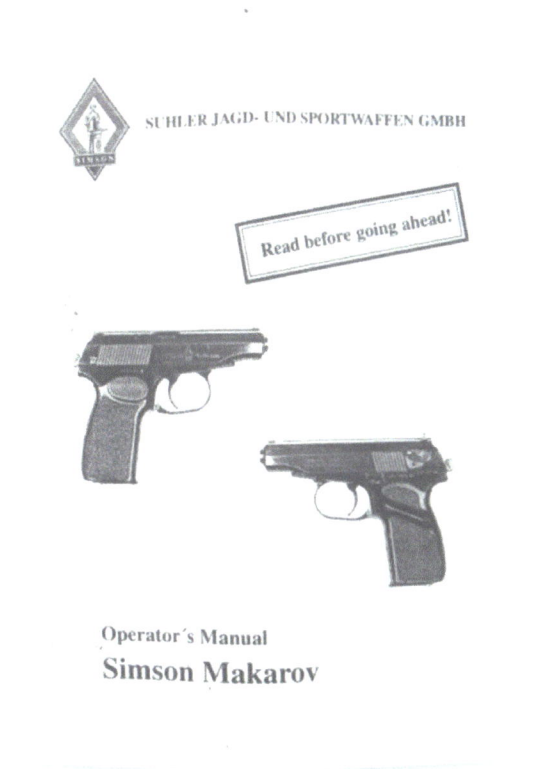

Simson Suhl Makarov operator's manual.[92]

Policeman of the Volkspolizei showing various firing positions with the Pistole M, from the Ministerium des Innern training manual, *Schießen mit den Pistolen Makarow und Modell 74.* Compare the prone position with the diagram from the Soviet Army manual on page 6.[93]

Front side of Volkspolizei log book or maintenance record for 1962 Pistole M number BS 4459. The log is marked 'BDVP Dresden' and 'StOV-WK VPKA'. This log identifies the issue of this particular Pistole M to the Volkspolizei Dresden District Office (Bezirksbehörde der Deutschen Volkspolizei Dresden) and to a specific municipal police station (VPKA - Volkspolizei Kreisämter).

Rear side of log book or maintenance record for 1962 Pistole M Number BS 4459.

Ephemera

Cutaway Patent Pistol

This cutaway patent model pistol is quite interesting and probably unique. The inventor, Mr. Roy L. Melcher, patented a magazine safety for the Makarov pistol, and a 1962 dated Pistole M was used as a model to show the modifications outlined in the patent. The patent was assigned to International Armament Corporation (Interarms) of Alexandria, Virginia. The modification, which added two parts (a small part to the right of the sear, and a torsion spring), but did not alter the original pistol in any way, changed the pistol so it could not be fired without the magazine inserted. Parts of the frame and slide of a Pistole M were cut to show this modification and use as a patent pistol. The details are found in U.S. patent number 5,388,362 which was granted on 14 February 1995.[94] It is not believed that this modification was ever made to any pistols on a production basis. The purpose of this modification was to make the pistol safer and undoubtedly to limit product liability.

United States Patent 5,388,362

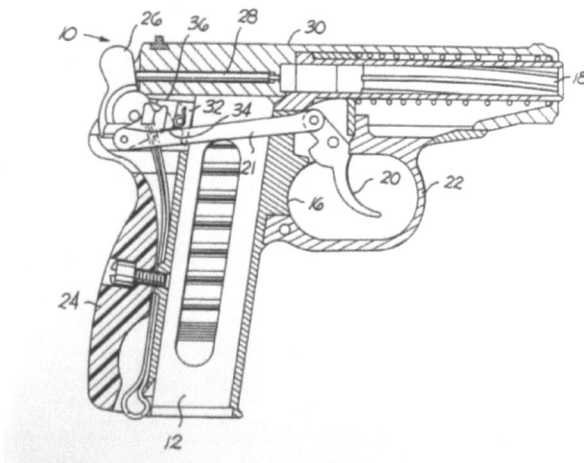

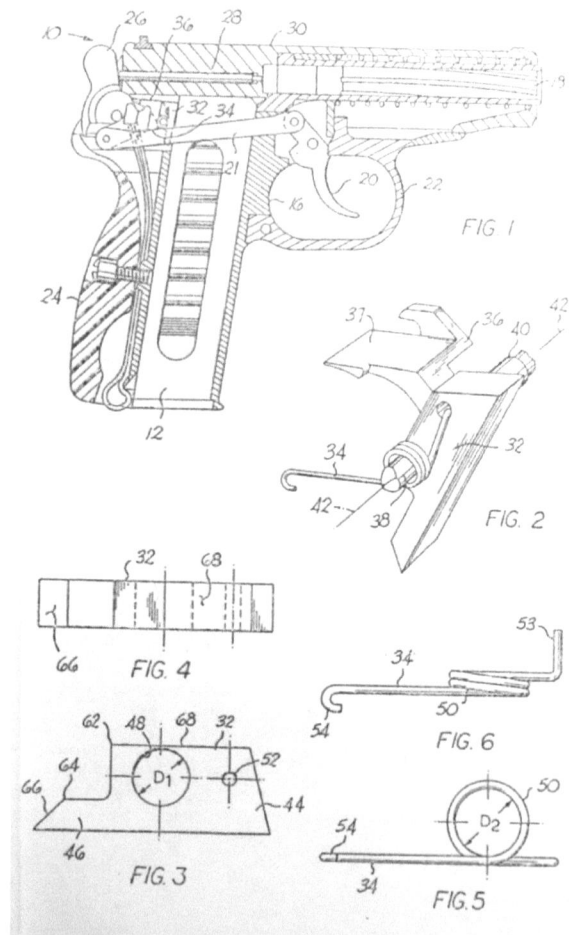

U.S. Patent: 5388362 - Magazine Safety for a Makarov Pistol.[95]

Right side of cutaway patent pistol. Note that the Pistole M has Bulgarian grips.

Close up of right side of cutaway patent pistol.

Close up of internal parts.

Importation into the United States and Import Markings

East German Makarovs were imported into the United States in large quantities the early to mid-1990s as military surplus. There were several importers, although it appears that Century Arms International imported the largest quantity. Most collectors were unfamiliar with the 9x18mm Makarov caliber when the pistols were first imported, so the Pistole M was not favored as much initially as other surplus pistols in more standard calibers. However, the pistols were priced very reasonably; between about $100 and $150, which made them affordable to many collectors. A review of an issue of *Shotgun News* from 1992 shows that they were listed at a price of $139 by at least two distributors.

Cover of Century International Arms Inc. sales flyer from the early 1990s, featuring the Pistole M. Note the reference to "STASI' Police' and the broken wall, a reference to fall of the Berlin Wall, in the background. Also note the thumb rest (target) grip which was required for importation.

The following is a listing of importers and their markings that are found on pistols imported into the United States.

International Arms Company (IACO) was based in Sacramento, California. IACO is reported to be one of the first importers of East German Makarovs. The number following the 'MM' marking appears to be the sequential number of the pistol imported. Based on these numbers, it appears that they imported several thousand pistols. The pistols have the following marking in very small font on the front of the grip strap:

ERNST THAELMANN MOD. M
9MM IACO SACCA GERMANY MMxxxx

IACO import marking located on the front grip strap of the frame. The number 3897 indicates this was the 3,897th pistol imported by IACO.

Century Arms International (CAI) was located in St. Albans, Vermont. The majority of pistols observed have CAI markings. The following markings are typically on the right side of the frame, though there are variations in the placement of the markings:

Germany CAI ST ALB VT
9mm MAK

CAI import marking with the word Germany stamped upside down. This may indicate that the two stampings were done at different times.

CAI import marking with caliber in the center.

CAI import marking with '9MM MAK' stamped at an angle.

Classic Distributers Incorporated (CDI) was based in Swanton, Vermont. CDI placed the following marking at the right rear of the frame in very small font:

Makarov 9 X 18 Germany
C.D.I. Swan. VT.

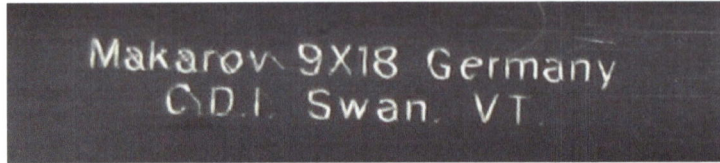

CDI import marking.

Cole Distributing is located in Smithville, Kentucky. Pistols imported by Cole Distributing are considered to be the best imports given the unobtrusiveness of the import marking. In addition, collectors like the fact that some of the pistols imported by Cole Distributing came with a matching East German maintenance log book. The import marking is stamped in very small font on the right side of the frame or on the bottom of the frame above the trigger area:

COLE DIST SVILLE KY

Cole import marking perpendicular to the frame directly behind the trigger guard.

SSME Deutsche Waffen Inc is located in Plant City, Florida. It appears that very few pistols were imported by SSME based on the low number of pistols seen with this marking. SSME imports have relatively small import markings. The markings are found on the right side of the frame above the trigger:

CAL 9x18
SSME PLANT CTY FL

SSME import marking.[96]

100

RUKO Products Incorporated was based in Buffalo, New York. RUKO imports are considered the least desirable based on their extensive import stamping, since many collectors feel that the amount of writing on the slide detracts from the appearance of the pistol. The lengthy three-line marking is found on the right side of the slide of some of the pistols:

Mod Makarov Cal9mm Makarov. Made in Germany by
Fahrzeug- u Jagdwaffenwerke Ernst- Thälmann, Suhl
Imported by RUKO Products Inc., Buffalo, NY

RUKO import marking with long text. Note that the text covers much of the side of the slide.

Other RUKO imported pistols have a shorter text marking on the right side of the slide.

Makarov Cal. 9x18   RUKO Products
Made By Ernst Thälmann   Buffalo, NY
Suhl, Germany

RUKO import marking with shorter text.

GP Trading was based in St. Albans, Vermont. GP Trading imports are marked on the rear of the right side of the frame in a small font:

Makarov mod. M 9 x 18 mm
Ernst Thaelmann. Germany
G.P. TRAD. St. Alb. Vt.

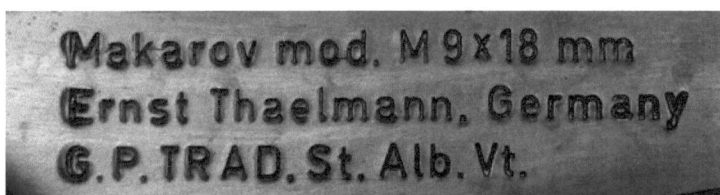

Long, but small and compact GP Trading import marking.

NHM of Sacramento, California is thought to stand for New Helvetia Mercantile which was part of the Old Sacramento Armoury. It appears that the number of pistols they imported was small based on how few have been observed. The pistols are marked in a relatively small font near the rear of the right side of the frame:

Makarov 9x18 Germany
N.H.M. SAC. CA

N.H.M. import marking.[97]

Import Grips

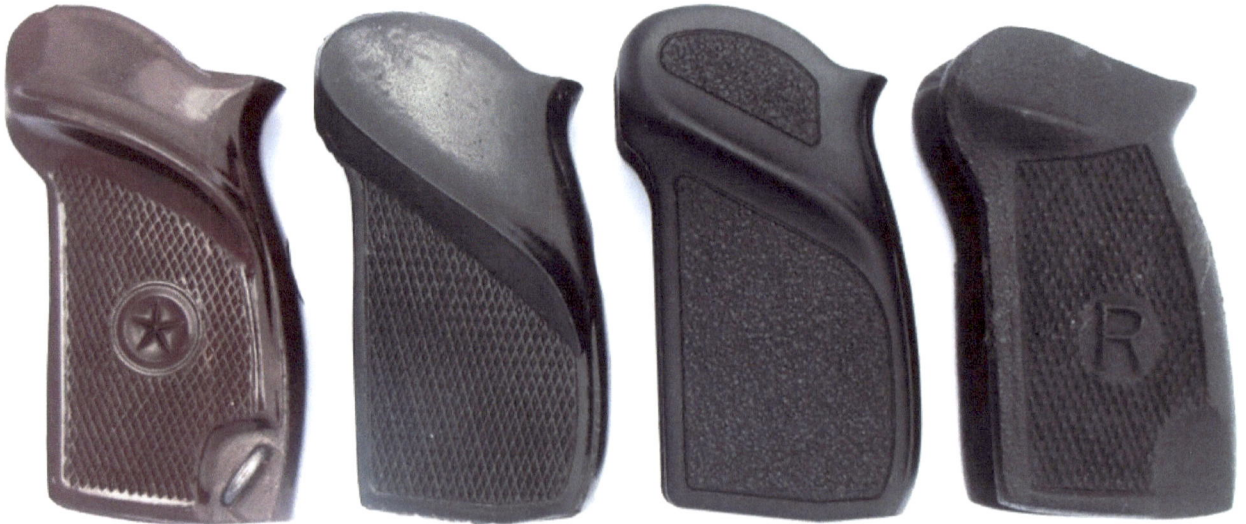

Grips put on pistols imported into the US. CAI pistols often have the brownish grips on the far left. RUKO pistols usually have the grip on the far right featuring a 'R'. Note that all have a thumb rest.

Import Boxes

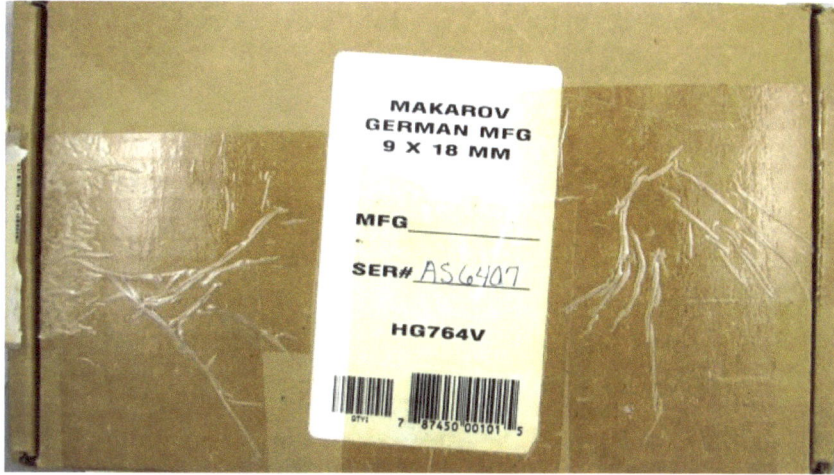

Cardboard box for pistol number AS 6407.

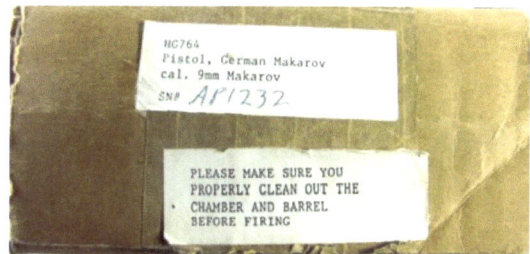

Opposite ends of a simple cardboard box for pistol AP1232.

Non-Import Marked Pistole M

It is also worth noting that there are a very few Pistole Ms in the United States that do not have import markings. There is a great deal of speculation as to how these pistols came to be in the United States. It is possible that they were brought in prior to the requirement that firearms have import markings or perhaps that were brought in discreetly by diplomats or by servicemen. However, without provenance, it is impossible to say how a given pistol was imported. These pistols are scarce and desirable since they have not been altered in any way and are in the most original form.

1965 Pistole M serial number FH 5444. Pistol has no import markings.

General Collector's Observations

Based on observations of a number of pistols over several years, a few general notes can be made.

Box: Some pistols imported by CAI featured a colorful box that had a picture of a Pistole M on the outside.

Factory re-work: One pistol observed had two sets of serial numbers on the slide and frame. The first set of numbers was barely visible under the finish and was in a slightly different spot than the second set of numbers. It is assumed that this was due to factory re-finishing. Perhaps the slide was replaced for some reason and the pistol was re-finished. The slide also had an upper case 'A' stamp on the inside.

Finish: Most Pistole Ms seem to have a deep black blue finish, but the finish appears to be more blue and 'commercial looking' on pistols dated 1964 and 1965. It is hard to say how the finish was applied, but it appears to be a rust blue. Another interesting observation is that the hammer and safety on some pistols is more of a plum color while others are a true blue or black color. The plum color is probably due to the composition of the steel which affected the final color of the finish.

Numbering on parts: A few internal parts (hammer, trigger bar, and sear) of some pistols have the last two digits of the serial number electro-pencilled. It is not known why this was done.

Value: The collector value of Makarov pistols have increased a great deal over time. The value has increased from $100 to $150 when first imported in the early 1990s, to $400-$600 in 2016. Selling prices at online auction rose substantially in 2014 and 2015 as collector interest increased. Prices are only expected to rise over time based on the fixed, limited supply and strong demand.

Relative Rarity: The below listing denotes the relative rarity of Pistole Ms in order of most to least rare.[98]

Scale: 1 = most rare, to 6 = least rare

Production Year	Relative Rarity	
1957 or 1958 prototype	no known examples	1
1959	perhaps 12 in the US	2
.22 Version	very few extant	2
Lehr Trainer	very few extant	2
No import marking	perhaps 50 in the US	3
Pistol with matching maintenance log	perhaps 50 in US	3
1960 to 1965 with matching magazine	perhaps 50 to 100 in US	4
Simson Suhl	several hundred in US	4
1963	less commonly observed year date	5
1960, 1961, 1962, 1964, 1965	commonly observed year dates	6

Serial numbers: The lowest number observed by collectors in a series is 0044 and the highest number is 98xx.

Sights: The rear sights often have different numbers that represent different heights.

Total number imported into the United States: Likely between 25,000 and 50,000.

Summary and Conclusion

The East German Makarov or Pistole M is an interesting and extremely high quality pistol that served the East Germany military and police for over thirty years. Several hundred thousand were produced at VEB Ernst Thälmann-Werk Suhl, East Germany from 1958 until October 1965. A large number were imported into the United States as military surplus in the early 1990s.

Many consider the Pistole M to be the best made Makarovs in terms of fit and finish. The level of quality is consistently high and is commensurate with commercially produced weapons. East German Makarovs are truly magnificent and alluring pistols.

Over time, the Pistole M has become a highly collectible and increasingly valued pistol. Some are drawn to it for its connection to the Cold War era, others like the quality and elegant design, some simply enjoy shooting them. Collectors have begun to recognize the subtle variations and nuances of the pistol. While we know much more about the Pistole M than ever before, there is much more to be learned. The legacy of the East German Makarov is that it serves as a tangible reminder of the Cold War era; a time and a place that will never be witnessed again.

Bild 27 Anschlag aus dem Kraftfahrzeug heraus

Policeman of the Volkspolizei demonstrating aiming the Pistole M from an automobile. From the Ministerium des Innern training manual, *Schießen mit den Pistolen Makarow und Modell 74.*[99]

Appendix

Makarov Pistol Timeline

1914	Nicolay Fyodorovich Makarov is born in Sasovo, Ryazan Oblast, Russian Empire.
1942	Izmech Factory 622 is established in Izhevsk, Urdmurtia Oblast, USSR.
1945	The Great Patriotic War ends. The search for a new generation of Soviet small arms is already underway.
1947	Initial batch of Makarov Pistol prototypes are manufactured at the Tula arsenal.
1949	Makarov Pistol production begins at Factory 622, Izhevsk for large scale trials.
1949	German Democratic Republic established, 7 October 1949.
1951	Soviet Union adopts the Makarov Pistol as its service pistol.
1953	A perfected pistol enters full scale production.
1957-1958	East German Pistole M pre-production period.
1959	Full licensed production of Pistole M begins at VEB Ernst Thälmann-Werk Suhl; breakage problems with investment cast parts.
1960	Pistole M parts fully machined starting around 1960.
1961	Soviet Makarov carried into space with cosmonaut Yuri Gagarin as part of the survival equipment on-board the spacecraft Vostok 1.
1965	Pistole M production ends, October 1965. The latest prefix observed by collectors is FJ.
1976	Investment cast parts begin to appear on Soviet produced Makarov pistols.
1988	Nicolay Fyodorovich Makarov passes away in Tula, USSR.
1990	German Democratic Republic dissolved, 3 October 1990. East Germany and West Germany reunify.
1991	The Union of Soviet Socialist Republics is dissolved, 26 December 1991. The Russian Federation becomes the successor state to the USSR. Izmech begins into a series of corporate reorganizations.
1992	Pistole Ms imported into the United States in large quantities.
1994	Simson Suhl Makarovs first produced. Importation of Soviet and Russian built Makarov pistols begins.
1996	Soviet and Russian Makarov imports are banned.
1997	Simson Suhl Makarovs sold in the United States.
2013	Kalashnikov Concern becomes the holding company producing Makarov pistols.

Makarov Disassembly Guide

The Makarov is easily disassembled using a procedure common to all Makarov pistols.

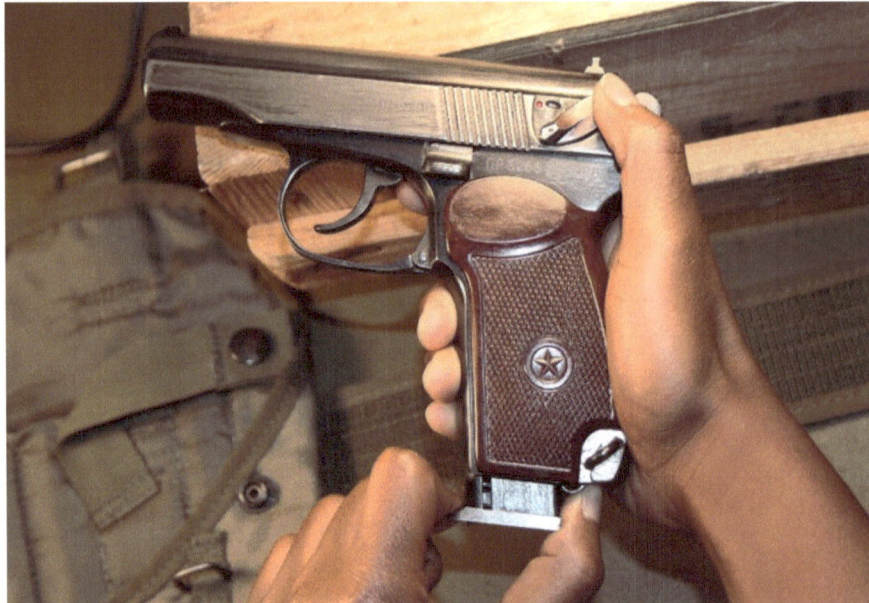

1. Point the pistol in a safe direction and keep fingers clear of the trigger. Apply the safety lever to the 'safe' position. Press rearward on the magazine catch and pull down on the magazine. Remove the magazine and place the safety lever in the 'fire' position.

2. Retract the slide to the rear to eject any chambered round

3.   Inspect the chamber to ensure that it is empty.

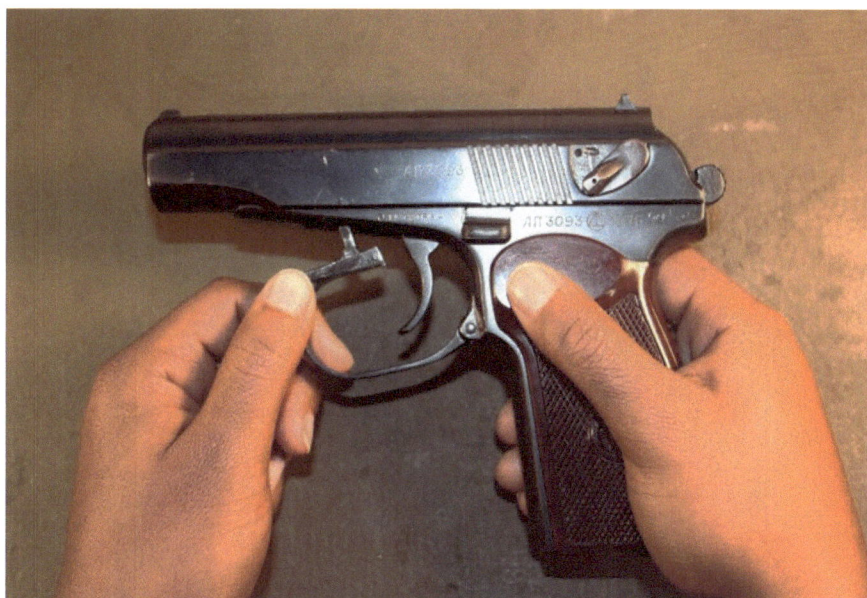

4.   Pull the front of the trigger guard down and hold it to one side using the index finger.

5. Retract the slide fully to the rear and lift it straight up and slide gently forward, so that it slides free over the barrel and off the frame.
    Note: This level of disassembly is considered sufficient for field stripping and is all that is necessary under normal circumstances.

6. Pull the recoil spring forward and simultaneously twist counter clockwise to remove. Note the orientation of the small end of the recoil spring at the chamber end of the barrel for re-assembly.

7. Lift the end of the sear spring off the back of the slide stop with the combination cleaning rod/tool, a screwdriver, or the base of the magazine.

8. Rotate the sear until the flat portion on the right side pivot pin aligns with the opening in the frame. Lift the slide stop/ejector up and out.

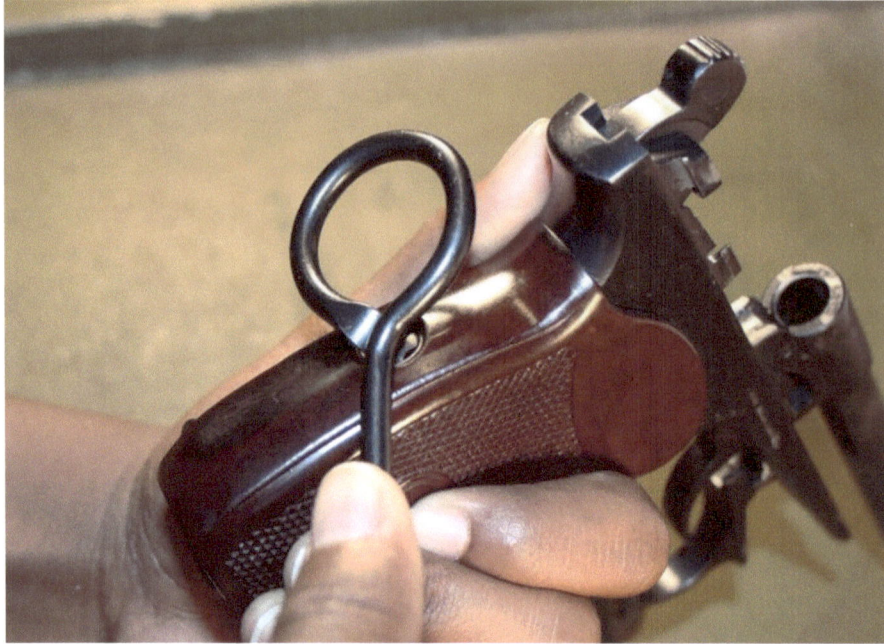

9. Remove the grip screw located in the rear of the grip.

10. Pull the grip to the rear and off the frame.

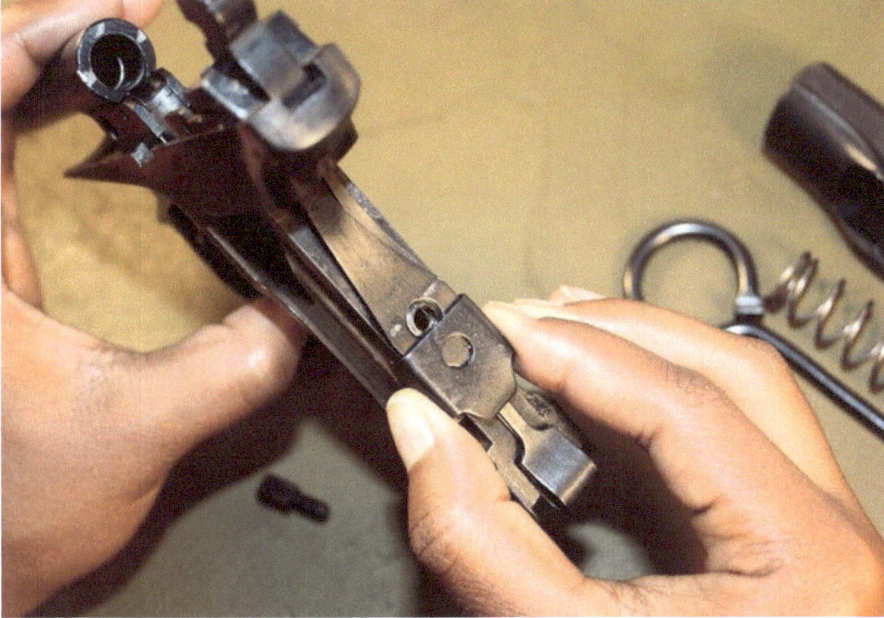

11. Slide the mainspring clamp down and off the frame. Remove the mainspring.

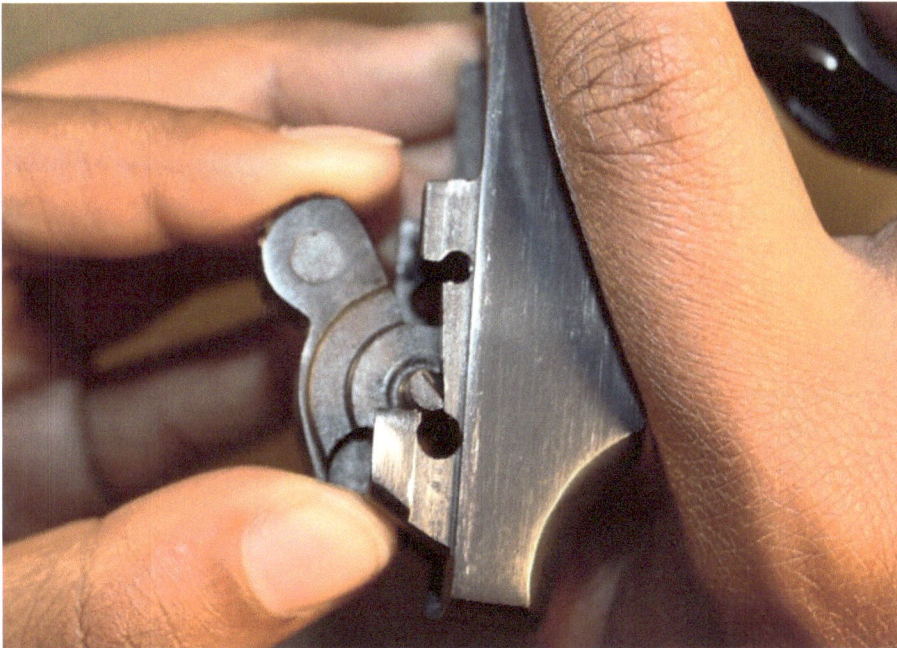

12. Rotate the hammer until the flat portions on the pivot pins line up with the openings in the frame and remove.

13. Remove the trigger bar by lifting it up and to the right of the frame.

14. Pull the trigger down and forward to remove from the frame.

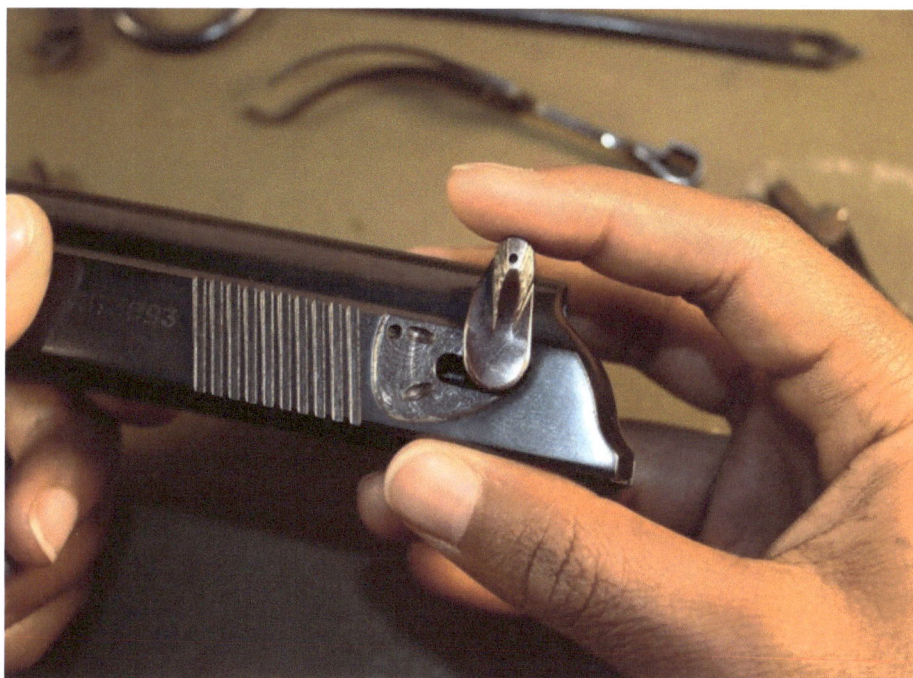

15. Rotate the safety lever up and over the slide and pull out. The firing pin will now drop out the rear of the slide. Note: before removing take time to examine how the rear of the firing pin is orientated in the slide.

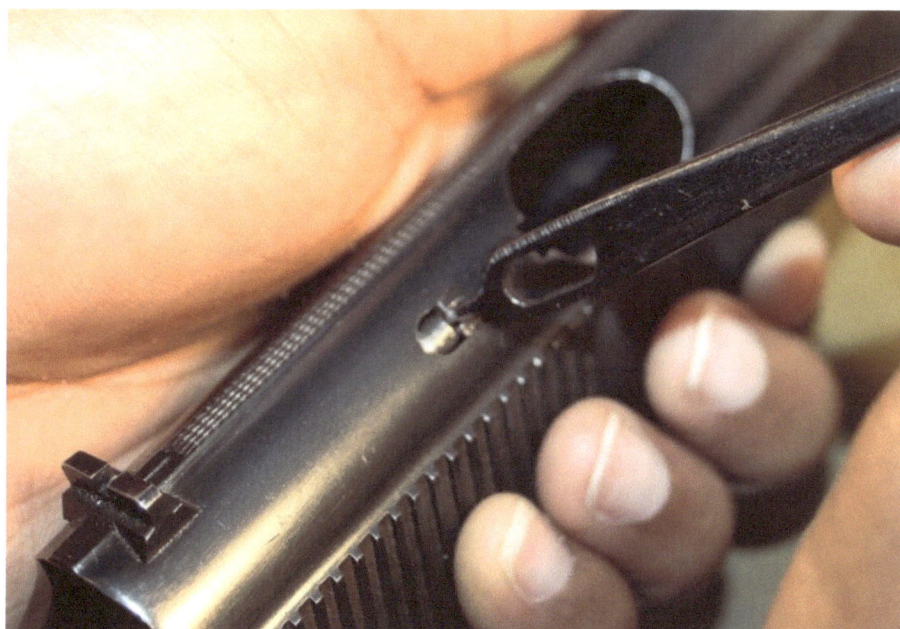

16. Find the small slot at the rear of the extractor. Using the tip of the cleaning rod/tool depress the spring loaded plunger towards the rear of the slide until the extractor can be rotated forward, down and out of its seat.

17. Reassemble in the reverse order. Sometimes the slide goes on more easily during reassembly by exerting light downward pressure on the slide when it is brought to the rear and put on the slide rails prior to it sliding forward.

Nicolay Fyodorovich Makarov (1914-1988)

Nicolay Fyodorovich Makarov (Николáй Фёдорович Макáров) was born in Sasovo, a town in the Ryazan Oblast, Russia in 1914. After successfully completing secondary school, he moved to Ryazan itself to attend a vocational school. After graduation he repaired steam engines at the Sasovo railway repair shop from 1931 to 1935. After further studies he was accepted into the Tula Mechanical Institute in 1936. Nicolay Makarov graduated in 1941 just in time for the 'Great Patriotic War' as Nazi Germany launched its invasion of the Soviet Union.

The wartime years saw Nicolay Makarov working in a factory producing the famous PPSh-41 submachine gun. Makarov eventually rose from foreman to chief designer in the plant and worked directly under the designer of the PPSh-41, Georgy Shpagin. In 1945 he was transferred to the design bureau in Tula where he headed a team that eventually designed his name sake, 'Makarov's Pistol', adopted in 1951 for the armed forces of the Soviet Union. His career went on from there, developing the A-12.7mm Afanazev aircraft machine gun design into the 23mm Afanasev-Makarov aircraft cannon adopted in 1956. From this point until his retirement he worked on the design of several rocket components.

For his life's work, Nicolay Makarov received numerous prestigious awards. These included the S.I. Mosin Prize, an Order of the Red Banner of Labor , two Orders of Lenin, two USSR State Prizes, and the title Hero of Socialist Labor. Despite these awards and other lesser ones, Nicolay Makarov has been often described as an extremely modest man who seldom spoke of his work outside of the office. He retired in 1974 and passed away in 1988 at the age of 74.[100]

Notes

Makarov: Soviet Union & Russian Federation

[1] *Armeyskiy Vestnik*, "*Kogda zamenyat pistolet Makarova?*", http://army-news.ru/2016/03/kogda-zamenyat-pistolet-makarova/

[2] Alfred Brückner and Gerhard Fischer, *Die Volkspolizei*, (Dresden:Verlag Zeit im Bild, 1975), 126.

[3] Alexander Korolov, "Pistol-packing for the Final Frontier: Why were Cosmonauts Armed?", *Russia Beyond the Headlines*, 17 July 2014.
http://rbth.com/defence/2014/07/17/pistol-packing_for_the_final_frontier_why_were_cosmonauts_armed_38279.html

[4] Ian V. Hogg, ed., *Jane's Infantry Weapons: 1988-89*, (Coulsdon: Jane's Information Group, 1988).

[5] Maxim Pyaduskin, Maria Houg and Anna Madoeva. "Beyond Kalashnikov: Small Arms Production, Exports, and Stockpiles in the Russian Federation", *Small Arms Survey Occasional Papers No. 10*, (2003).
http://www.smallarmssurvey.org/fileadmin/docs/B-occasional-papers/SAS-OP10-Russia.pdf

[6] *Ibid.*

[7] Makarov.com (2016), http://www.makarov.com

[8] Sergey Kuznetsov in Oleg Korjakin, "*Ispytaniya pistoleta budushchego "Udav" zavershatsya v 2016 godu*", *Rossiyskoy Gazety*, 15 June 2016, https://rg.ru/2016/06/15/ispytaniia-pistoleta-budushchego-udav-zavershatsia-v-2016-godu.html

[9] David N. Bolotin, *Military History Library: History of Soviet Small Arms and Ammunition*, (St. Petersburg: Polygon, 1995), 24

[10] Elena M. Kalashnikova, *Kalashnikov: The Inside Story of the Designer and His Weapons*, (Ironsides: Ironside International Publishers, 2009), 156.

[11] Bolotin, *Military History Library: History of Soviet Small Arms and Ammunition*, 24

[12] *Ibid.*, 24

[13] Fred A Datig, *The History and Development of Imperial and Soviet Russian Military Small Arms and Ammunition 1700-1986. Volume Thirteen: Soviet Russian Tokarev "TT" Pistols and Cartridges 1929-1953*. (Culver City, CA: Michael Zomber Co., 1990), 89.

[14] *Everything About the Makarov Pistol, PM*, Yucoz 2, 2016 6 January 2016, http://pm9.ucoz.ru

[15] S.B. Monetchikov, *Domestic Revolvers and Pistols*, (St. Petersburg: Atlas Publishing House, 2015), 250.

[16] *Everything About the Makarov Pistol, PM*, *op. cit.*

[17] Ministry of Defense of the Union of Soviet Socialists Republics, *Manual of Instructions for Use and Maintenance for 9mm Makarov Pistol*, 2nd. ed., (Moscow: Military Publishing House of the Ministry of Defense of the Union of Soviet Socialists Republics, 1955), 74

[18] "Evolution", *Everything About the Makarov Pistol, PM*, http://pm9.ucoz.ru/index/ehvoljucija_pm/0-63

[19] Monetchikov, *op. cit.*, 255

[20] Ministry of Defense of the Union of Soviet Socialists Republics, *9mm Makarov Pistol (PM) Handbook for Repairs*, (Moscow: Publishing House of the Ministry of Defense of the Union of Soviet Socialists Republics, 1956)

[21] *Manual of Instructions for Use and Maintenance for 9mm Makarov Pistol*, 2nd. ed., *op.cit.*

[22] Photo: Andrey Danko

[23] Photo: Andrey Danko

[24] "Evolution", *Everything About the Makarov Pistol, PM*, *op. cit.*

[25] Ministry of Defense of the Union of Soviet Socialists Republics, *9mm Makarov Pistol (PM) Handbook for Repairs*, *op. cit.*

[26] Monetchikov, *op. cit.*, 255

[27] *Manual of Instructions for Use and Maintenance for 9mm Makarov Pistol*, (Moscow: Military Publishing House,1986), 6

[28] Fred A Datig, *The History and Development of Imperial and Soviet Russian Military Small Arms and Ammunition 1700-1986, Volume Sixteen: Soviet Russian Postwar Military Pistols and Cartridges 1945-1986*, (Glenview, IL: Handgun Press, 1988), 55.

[29] *Ibid.*, 53

[30] *Everything About the Makarov Pistol, PM*, Yucoz 2, 2016 6 January 2016, http://pm9.ucoz.ru

[31] *Manual of Instructions for Use and Maintenance for 9mm Makarov Pistol*, (Moscow: Military Publishing House,1986), 45

[32] Monetchikov, *op. cit.*, 118

[33] Hogg, ed., *op.cit.*, 57

[34] Giacomo P. Paoli, "The Method Behind the Mark: A Review of Firearm Marking Technologies", *Small Arms Survey Issue Brief*, No. 1, Dec. 2010, http://www.smallarmssurvey.org/fileadmin/docs/G-Issue-Brieds/SAS_IBI_Method-Behind-The-Mark.pdf

[35] Benjamin King, ed., "Surveying European Production and Procurement of Small Arms and Light Weapons Ammunition: The Cases of Italy, France, and the Russian Federation", *Small Arms Survey Occasional Paper No, 10*, (Geneva: Graduate Institute of International Studies, 2010).
http://www.smallarmssurvey.org/fileadmin/docs/B-occasional-papers/SAS-OP10-Production.pdf

[36] Photo: Aldali

[37] "About", *Kalashnikov Concern*, http://kalashnikovconcern.ru/en/about/

[38] Monetchikov, *op. cit.*, 260

[39] Photo: popovaphoto

[40] Monetchikov, *op. cit.*, 446

[41] "About", *Kalashnikov Concern*, *op.cit.*

[42] Paoli, *op.cit.*

[43] Photo: Brian J. Driessen

[44] George Mellinger, *Osprey Aircraft of the Aces No. 64: Yakovlev Aces of World War 2*, (Botley: Osprey Publishing, 2005).

[45] Charles Ward, "Baikal MP 654K Special Edition, Makarov", *Replica CO2 Air Pistol Collection*, (2 February 2016). http://www.Co2Airguns.net/collection/Baikal/Makarov/index.htm

[46] Photo: Cameron S. White

[47] *Ibid.*

[48] Kalashnikov Concern, *Products: Military Law Enforcement*; Kalashnikov Concern JSC. (2016). http://www.kalashnikovconcern.ru/en/kalashnikov/products/mle

[49] *Manual of Instructions for Use and Maintenance for 9mm Makarov Pistol, op.cit.*, 45

[50] Photo: Bork

[51] Datig, *Soviet Russian Postwar Military Pistols and Cartridges 1945-1986, op.cit.*, 82

[52] Bolotin, *op.cit.*, 26

[53] Photo: zim90

[54] Photo: mrakor

[55] Bolotin, *op.cit.*, 33

[56] Ibid., 76, 71

[57] Photos: www.flora-sakura.com.ua

[58] Makarov.com (2016), *op.cit.*

Pistole M: East Germany

[59] All power to the Soviets!, http://partisan1943.tumblr.com/post/134748775305/soldier-of-the-east-german-national-people-s-army

[60] The East German Makarov pistol has been referred to various ways in the literature. The terms 'East German Makarov', '9mm-Pistole M' and 'Pistole M' (abbreviated as 'PM') are used interchangeably. In this book, they refer specifically to the East German manufactured Makarov. The pistol has also been referred to less commonly as the 'PM9'.

[61] Stasi is the abbreviation of Staatssicherhit, or State Security Service. The East German population was terrified of the Stasi who kept extensive records on people and who terrorized them with subtle intimidation and overt threats and abuse.

[62] The *Deutsche Demokratische Republik (DDR)* or German Democratic Republic (GDR) is commonly referred to as 'East Germany'.

[63] *Sozialistische Einheitspartei Deutschlands* (SED)

[64] 52 Pickup, *Based on map data of the IEG-Maps project (Andreas Kunz, B. Johnen and Joachim Robert Moeschl: University of Mainz) - www.ieg-maps.uni-mainz.de.*, https://en.wikipedia.org/wiki/Allied-occupied_Germany#/media/File:Map-Germany-1945.svg

[65] Dieter H. Marschall, *Handguns of the Armed Organizations of the Soviet Occupation Zone and the German Democratic Republic*, 1st ed. English language, (Los Alamos: Ucross Books, 2003).

[66] *Ibid.*, 25

[67] *Makarov Operator Manual*, Federal People's Army of the German Democratic Republic A 255/1/109, 1972, English translation by John Baum, 2011.

[68] Ernst Thälmann (1886-1944) was a German Communist Party leader in the early part of the 20th century. He was imprisoned during the 1930s and perished in the Nazi concentration camp at Buchenwald. Many building in East German were named in honor of this famous leader. VEB is the abbreviation of *Volkseigener Betrieb*, or an East German state owned company.

[69] "Pistol Makarov 9 mm - Manufacturer: Armories Suhl, Germany", www.makarov.ch, http://www.makarov.ch/index.php?page=suhl

[70] Metilsteiner, *District-Map of Suhl in Thuringia*, https://en.wikipedia.org/wiki/File:Suhl_Stadtgliederung.png.

[71] See Edward B. Tinker and Graham K. Johnson, *Simson Lugers Simson & Co, Suhl, the Weimar Years*, (Galesburg, IL: Brad Simpson Publishing, 2007) for a complete history of the Simson firm; also see German language works Siegfried Schutt, *Die Simson-Legende Aus der Geschichte eines Traditionsunterrnehmens*, (Druckmedienzentrum, 2006), and Ulrike Von Schulz, *Simson: Vom unwahrseheimlichen Uberleben eines Unternehmens 1856-1993*, (Wallstein, 2013) for additional information on the history of the Simson firm.

[72] Marschall, *op.cit.*,19

[73] Wilfried Kopenhagen, *Die Mot-Schutzen der NVA*, (Solingen: Barett Verlag GmbH, 1995).

[74] Marschall, *op.cit.*, 35

[75] Data from observation of collector's samples by the author, and by collectors, listed on www.Makarov.com.

[76] *Schießen mit den Pistolen Makarow und Modell 74*, (Berlin: Ministerium des Innern Publikationsabteilung, 1986).

[77] Reported by www.Makarov.com, Marschall *op.cit.*, Edward Clinton Ezell, *Small Arms of the World*, 12th Revised Edition. (Harrisburg: Stackpole Books, 1983).

[78] Photo: Edward Clinton Ezell, *Small Arms of the World*, 12th Revised Edition. (Harrisburg: Stackpole Books, 1983).

[79] Photo: Roger A. Finzel

[80] Toni02, "Waffen im MfS, was es alles gab...", 09.November 2010 15:55, *NVA Forum.* http://www.nva-forum.de/nva-board/index.php?showtopic=11520&view=findpost&p=385931

[81] *Handbuch für den Grenzdienst*, 5. Aufl. (Berlin:Ministerium für Nationale Verteidigung der Deutschen Demokratischen Republik, 1978), 122. Photo:Garandfan43, http://forums.gunboards.com/attachment.php?attachmentid=650879&d=1364756189

[82] *Schießen mit den Pistolen Makarow und Modell 74*, (Berlin: Ministerium des Innern Publikationsabteilung, 1986), 28

[83] Photo: Edwin H. Lowe

[84] Marschall, *op.cit.*, 20, 23

[85] *Ibid.*, 1

[86] *Ibid.*, 53

[87] *9-mm Pistole Makarow: Beschreibung und Nutzung A 250/1/109*, (Berlin: Nationale Volksarmee der Deutschen Demokratischen Republik, 1975)

[88] *Ibid.*

[89] *9-mm-Pistole M: Prüftechnologie 050/1/442* (Berlin: Nationale Volksarmee der Deutschen Demokratischen Republik, 1976)

[90] *9-mm-Pistole M: Befundaufnahme und Qualitätsfeststellung 050/1/442* (Berlin: Nationale Volksarmee der Deutschen Demokratischen Republik, 1976)

[91] *Schießen mit den Pistolen Makarow und Modell 74*, (Berlin: Ministerium des Innern Publikationsabteilung, 1986)

[92] Operator's Manual: Simson Makarov, (Suhl: Suhler Jagd-und Sportwaffen GmBH)

[93] *Schießen mit den Pistolen Makarow und Modell 74*, *op.cit.*

[94] Roy L. Melcher, "U.S. Patent: 5388362 - Magazine Safety for a Makarov Pistol", 14 February 1995.

[95] *Ibid.*

[96] Photo: J. Shank. Pistol from T. Jevicky collection

[97] Photo: J. Shank

[98] Based on JamesD, "East German Makarov 'Pecking Order'", forum thread, 1-9-2016, www.Makarov.com.

[99] *Schießen mit den Pistolen Makarow und Modell 74*, *op.cit.*

[100] Bolotin, *op.cit.*, 263

Bibliography

Works Cited or Reviewed

Barnes, Frank C., *Cartridges of the World, 6th Edition*, (DBI Books, 1989).

Bolotin, David N., *Military History Library: History of Soviet Small Arms and Ammunition*, (St. Petersburg: Polygon, 1995).

Bolotin, David N., *Soviet Small-Arms and Ammunition*, English translation, edited by John Walter and Heikki Pohjolainen, (Helsinki: Finnish Arms Museum Foundation, 1995).

Combat Pistols of Russia: TT. Makarov. PSM. Stechkin. Know What You Own Series, (Moscow: Geleos Publishing House, 2003).

Cude, Walton H., *The Ultimate Makarov Guide The Complete Owner's Manual.* (Cude Gunworks/Gundata Publications, 1996).

Datig, Fred A., *The History and Development of Imperial and Soviet Russian Military Small Arms and Ammunition 1700-1986, Volume Thirteen: Soviet Russian Tokarev "TT" Pistols and Cartridges 1929-1953*, (Culver City, CA: Michael Zomber Co., 1990).

Datig, Fred A., *The History and Development of Imperial and Soviet Russian Military Small Arms and Ammunition 1700-1986, Volume Sixteen: Soviet Russian Postwar Military Pistols and Cartridges 1945-1986*, (Glenview, IL: Handgun Press, 1988).

Desmond, Dennis, *Camouflage Uniforms of the Soviet Union and Russia: 1937-to the Present*, (Atglen, PA: Schiffer Publishing Ltd., 1998).

Everything About the Makarov Pistol, PM, Yucoz 2, 6 January 2016, http://pm9.ucoz.ru

Ezell, Edward C., *Handguns of the World Military Revolvers and Self-Loaders from 1870 to 1945*, (Harrisburg: Stackpole Books, 1981).

Ezell, Edward Clinton., *Small Arms of the World*, 12th Revised Edition. (Harrisburg: Stackpole Books, 1983).

Firearms Assembly 4: The NRA Guide to Pistols and Revolvers. (NRA Book Service, 1980).

Hogg, Ian V. ed., *Jane's Infantry Weapons: 1988-89*, (Coulsdon, Surrey: Jane's Information Group, 1988).

Identification Guide Military Small Calibre Ammunition (Up to 15mm) (For Forensic and Law Enforcement Use, 1989 Edition, (Rhino Publishing Company, 1989).

Kalashnikov Concern, http://www.kalashnikovconcern.ru/en/kalashnikov/products/mle

Kalashnikova, Elena M., *Kalashnikov: The Inside Story of the Designer and His Weapons*, (Ironsides, MD: Ironside International Publishers, 2009).

King, Benjamin, ed., "Surveying European Production and Procurement of small Arms and Light Weapons Ammunition: The Cases of Italy, France, and the Russian Federation". *Small Arms Survey Occasional Paper No, 10.* (Geneva: Graduate Institute of International Studies, 2010). http://www.smallarmssurvey.org/fileadmin/docs/B-occasional-papers/SAS-OP10-Production.pdf

Kokalis, Peter G., "Makarov: Pipsqueak or Concealable Powerhouse?", *Shotgun News Treasury #7*, pp. 92-99.

Kopenhagen, Wilfried., *Die Mot-Schutzen der NVA*, (Solingen: Barett Verlag GmbH, 1995).

Korolov, Alexander, "Pistol-packing for the Final Frontier: Why were Cosmonauts Armed?", *Russia Beyond the Headlines*, 17 July, 2014, http://www.rbth.com

Lawrence, Erik., *Practical Guide to the Operational Use of the Makarov Pistol, 2nd Edition*, (2014)

Makarov.ch. http://www.Makarov.ch

Makarov.com. http://www.Makarov.com

Makarov Operator Manual, Federal People's Army of the German Democratic Republic A 255/1/109, 1972, English translation by John Baum, 2011.

Manual of Instructions for Use and Maintenance for 9mm Makarov Pistol, (Moscow: Military Publishing House,1986),

Marschall, Dieter H., *Handguns of the Armed Organizations of the Soviet Occupation Zone and German Democratic Republic,* (Los Alamos, NM: Ucross Books, 2003).

Mellinger, George, *Osprey Aircraft of the Aces No. 64: Yakovlev Aces of World War 2,* (Botley: Osprey Publishing, 2005).

Ministry of Defense of the Union of Soviet Socialists Republics, *9mm Makarov Pistol (PM) Handbook for Repairs,* (Moscow: Publishing House of the Ministry of Defense of the Union of Soviet Socialists Republics, 1956)

Ministry of Defense of the Union of Soviet Socialists Republics, *Manual of Instructions for Use and Maintenance for 9mm Makarov Pistol,* 2nd. ed., (Moscow: Military Publishing House of the Ministry of Defense of the Union of Soviet Socialists Republics, 1955)

Ministerium des Innern, *Schießen mit den Pistolen Makarov und Modell 74,* (Berlin: Ministerium des Innern Publikationsabteilung, 1986).

Ministerium für Nationale Verteidigung der Deutschen Demokratischen Republik, *Handbuch für den Grenzdienst,* 5. Aufl. (Berlin:Ministerium für Nationale Verteidigung der Deutschen Demokratischen Republik, 1978)

Monetchikov, S.B., *Domestic Revolvers and Pistols,* (St. Petersburg: Atlas Publishing House, 2015).

Paoli, Giacomo P., "The Method Behind the Mark: A Review of Firearm Marking Technologies". *Small Arms Survey Issue Brief* No. 1, Dec. 2010. http://www.smallarmssurvey.org/fileadmin/docs/G-Issue-Brieds/SAS_IBI_Method-Behind-The-Mark.pdf

Pearce, Lane., "Century Czech Model 52, 7.62 Tokarev, German Makarov 9mm". *Handgunning,* March/April 1993, pp.26-27.

Pyaduskin, Maxim., Houg, Maria. and Madoeva, Anna., "Beyond Kalashnikov: Small Arms Production, Exports, and Stockpiles in the Russian Federation", *Small Arms Survey Occasional Papers No. 10,* (2003). http://www.smallarmssurvey.org/fileadmin/docs/B-occasional-papers/SAS-OP10-Russia.pdf

Schutt, Siegfried, *Die Simson-Legende Aus der Geschichte eines Traditionsunterrnehmens,* (Druckmedienzentrum, 2006).

Smith, W.H.B., *Small Arms of the World: A Basic Manual of Small Arms. 10th ed.,* (New York: Galahad Books, 1973).

Shimek, Robert T., "Makarov". *Surplus Firearms, Volume III,* 1994, pp. 80-85.

Skennerton, Ian D., *9mm Makarov Pistol Handbook for Identification Stripping & Assembly Service & Maintenance Operation & Function Illustrated Parts Lists Accessories & Spares Historical Memoranda Additional References,* (Ray Riling Arms Books, 2005).

Tinker, Edward B. and Johnson, Graham K., *Simson Lugers Simson & Co, Suhl, the Weimar Years,* (2007).

Von Schulz, Ulrike., *Simson: Vom unwahrseheimlichen Uberleben eines Unternehmens 1856-1993,* (Wallstein, 2013).

Ward, Charles., "Baikal MP 654K Special Edition, Makarov", *Replica Co2 Air Pistol Collection,* (2 February 2016). http://www.Co2Airguns.net/collection/Baikal/Makarov/index.htm

White, Cameron S., "The East German Makarov: A Cold War Classic". *American Rifleman,* August 2015, pp. 66-69, 94, 99.

www.Wikipedia.com, History of East Germany

www.Wikipedia.com, Makarov.

www.Wikipedia.com, Thaelmann, Ernst.

Index